The Elements of Geology

by W. H. Norton

PREFACE

Geology is a science of such rapid growth that no apology is expected when from time to time a new text-book is added to those already in the field. The present work, however, is the outcome of the need of a text-book of very simple outline, in which causes and their consequences should be knit together as closely as possible,--a need long felt by the author in his teaching, and perhaps by other teachers also. The author has ventured, therefore, to depart from the common usage which subdivides geology into a number of departments,--dynamical, structural, physiographic, and historical,--and to treat in immediate connection with each geological process the land forms and the rock structures which it has produced.

It is hoped that the facts of geology and the inferences drawn from them have been so presented as to afford an efficient discipline in inductive reasoning. Typical examples have been used to introduce many topics, and it has been the author's aim to give due proportion to both the wide generalizations of our science and to the concrete facts on which they rest.

There have been added a number of practical exercises such as the author has used for several years in the class room. These are not made so numerous as to displace the problems which no doubt many teachers prefer to have their pupils solve impromptu during the recitation, but may, it is hoped, suggest their use.

In historical geology a broad view is given of the development of the North American continent and the evolution of life upon the planet. Only the leading types of plants and animals are mentioned, and special attention is given to those which mark the lines of descent of forms now living.

By omitting much technical detail of a mineralogical and paleontological nature, and by confining the field of view almost wholly to our own continent, space has been obtained to give to what are deemed for beginners the essentials of the science a fuller treatment than perhaps is common.

It is assumed that field work will be introduced with the commencement of the study. The common rocks are therefore briefly described in the opening chapters. The drift also receives early mention, and teachers in the northern states who begin geology in the fall may prefer to take up the chapter on the Pleistocene immediately after the chapter on glaciers.

Simple diagrams have been used freely, not only because they are often clearer than any verbal statement, but also because they readily lend themselves to reproduction on the blackboard by the pupil. The text will suggest others which the pupil may invent. It is hoped that the photographic views may also be used for exercises in the class room.

The generous aid of many friends is recognized with special pleasure. To Professor W. M. Davis of Harvard University there is owing a large obligation for the broad conceptions and luminous statements of geologic facts and principles with which he has enriched the literature of our science, and for his stimulating influence in education. It is hoped that both in subject-matter and in method the book itself makes evident this debt. But besides a general obligation shared by geologists everywhere, and in varying degrees by perhaps all authors of recent American text- books in earth science, there is owing a debt direct and personal. The plan of the book, with its use of problems and treatment of land forms and rock structures in immediate connection with the processes which produce them, was submitted to Professor Davis, and, receiving his approval, was carried into effect, although without the sanction of precedent at the time. Professor Davis also kindly consented to read the manuscript throughout, and his many helpful criticisms and suggestions are acknowledged with sincere gratitude.

Parts of the manuscript have been reviewed by Dr. Samuel Calvin and Dr. Frank M. Wilder of the State University of Iowa; Dr. S. W. Beyer of the Iowa College of Agriculture and Mechanic Arts; Dr. U. S. Grant of Northwestern University; Professor J. A. Udden of Augustana College, Illinois; Dr. C. H. Gordon of the New Mexico State School of Mines; Principal Maurice Ricker of the High School, Burlington, Iowa; and the following former students of the author who are engaged in the earth sciences: Dr. W. C. Alden of the United States Geological Survey and the University of Chicago; Mr. Joseph Sniffen, instructor in the Academy of the University of Chicago, Morgan Park; Professor Martin Iorns, Fort Worth University, Texas; Professor A. M. Jayne, Dakota University; Professor G. H. Bretnall, Monmouth College, Illinois; Professor Howard E. Simpson, Colby College, Maine; Mr. E. J. Cable, instructor in the Iowa State Normal College; Principal C. C. Gray of the High

School, Fargo, North Dakota; and Mr. Charles Persons of the High School, Hannibal, Missouri. A large number of the diagrams of the book were drawn by Mr. W. W. White of the Art School of Cornell College. To all these friends, and to the many who have kindly supplied the illustrations of the text, whose names are mentioned in an appended list, the writer returns his heartfelt thanks.

WILLIAM HARMON NORTON

CORNELL COLLEGE, MOUNT VERNON, IOWA

JULY, 1905

INTRODUCTION.--THE SCOPE AND AIM OF GEOLOGY

PART I

EXTERNAL GEOLOGICAL AGENCIES

I. THE WORK OF THE WEATHER II. THE WORK OF GROUND WATER III. RIVERS AND VALLEYS IV. RIVER DEPOSITS V. THE WORK OF GLACIERS VI. THE WORK OF THE WIND VII. THE SEA AND ITS SHORES VIII. OFFSHORE AND DEEP-SEA DEPOSITS

PART II

INTERNAL GEOLOGICAL AGENCIES

IX. MOVEMENTS OF THE EARTH'S CRUST X. EARTHQUAKES XI. VOLCANOES XII. UNDERGROUND STRUCTURES OF IGNEOUS ORIGIN XIII. METAMORPHISM AND MINERAL VEINS

PART III

HISTORICAL GEOLOGY

XIV. THE GEOLOGICAL RECORD XV. THE PRE-CAMBRIAN SYSTEMS XVI. THE CAMBRIAN XVII. THE ORDOVICIAN AND SILURIAN XVIII. THE DEVONIAN XIX. THE CARBONIFEROUS XX. THE MESOZOIC XXI. THE TERTIARY XXII. THE QUATERNARY INDEX

THE ELEMENTS OF GEOLOGY

INTRODUCTION

THE SCOPE AND AIM OF GEOLOGY

Geology deals with the rocks of the earth's crust. It learns from their composition and structure how the rocks were made and how they have been modified. It ascertains how they have been brought to their present places and wrought to their various topographic forms, such as hills and valleys, plains and mountains. It studies the vestiges which the rocks preserve of ancient organisms which once inhabited our planet. Geology is the history of the earth and its inhabitants, as read in the rocks of the earth's crust.

To obtain a general idea of the nature and method of our science before beginning its study in detail, we may visit some valley, such as that illustrated in the frontispiece, on whose sides are rocky ledges. Here the rocks lie in horizontal layers. Although only their edges are exposed, we may infer that these layers run into the upland on either side and underlie the entire district; they are part of the foundation of solid rock which everywhere is found beneath the loose materials of the surface.

The ledges of the valley of our illustration are of sandstone. Looking closely at the rock we see that it is composed of myriads of grains of sand cemented together. These grains have been worn and rounded. They are sorted also, those of each layer being about of a size. By some means they have been brought hither from some more ancient source. Surely these grains have had a history before they here found a resting place,--a history which we are to learn to read.

The successive layers of the rock suggest that they were built one after another from the bottom upward. We may be as sure that each layer was formed before those above it as that the bottom courses of stone in a wall were laid before the courses which rest upon them.

We have no reason to believe that the lowest layers which we see here were the earliest ever formed. Indeed, some deep boring in the vicinity may prove that the ledges rest upon other layers of rock which extend downward for many hundreds of feet below the valley floor. Nor may we conclude that the highest layers here were the latest ever laid; for elsewhere we may find still later layers lying upon them.

A short search may find in the rock relics of animals, such as the imprints of shells, which lived when it was deposited; and as these are of kinds whose nearest living relatives now have their home in the sea, we infer that it was on the flat sea floor that the sandstone was laid. Its present position hundreds of feet above sea level proves that it has since emerged to form part of the land;

while the flatness of the beds shows that the movement was so uniform and gentle as not to break or strongly bend them from their original attitude.

The surface of some of these layers is ripple-marked. Hence the sand must once have been as loose as that of shallow sea bottoms and sea beaches to-day, which is thrown into similar ripples by movements of the water. In some way the grains have since become cemented into firm rock.

Note that the layers on one side of the valley agree with those on the other, each matching the one opposite at the same level. Once they were continuous across the valley. Where the valley now is was once a continuous upland built of horizontal layers; the layers now show their edges, or OUTCROP, on the valley sides because they have been cut by the valley trench.

The rock of the ledges is crumbling away. At the foot of each step of rock lie fragments which have fallen. Thus the valley is slowly widening. It has been narrower in the past; it will be wider in the future.

Through the valley runs a stream. The waters of rains which have fallen on the upper parts of the stream's basin are now on their way to the river and the sea. Rock fragments and grains of sand creeping down the valley slopes come within reach of the stream and are washed along by the running water. Here and there they lodge for a time in banks of sand and gravel, but sooner or later they are taken up again and carried on. The grains of sand which were brought from some ancient source to form these rocks are on their way to some new goal. As they are washed along the rocky bed of the stream they slowly rasp and wear it deeper. The valley will be deeper in the future; it has been less deep in the past.

In this little valley we see slow changes now in progress. We find also in the composition, the structure, and the attitude of the rocks, and the land forms to which they have been sculptured, the record of a long succession of past changes involving the origin of sand grains and their gathering and deposit upon the bottom of some ancient sea, the cementation of their layers into solid rock, the uplift of the rocks to form a land surface, and, last of all, the carving of a valley in the upland. Everywhere, in the fields, along the river, among the mountains, by the seashore, and in the desert, we may discover slow changes now in progress and the record of similar changes in the past. Everywhere we may catch glimpses of a process of gradual change, which stretches backward into the past and forward into the future, by which the forms and structures of the face of the earth are continually built and continually destroyed. The science which deals with this long process is geology. Geology treats of the natural changes now taking place upon the earth and within it, the agencies which produce them, and the land forms and rock structures which result. It studies the changes of the present in order to be able to read the history of the earth's changes in the past.

The various agencies which have fashioned the face of the earth may. be divided into two general classes. In

Part I we shall

consider those which work upon the earth from without, such as the weather, running water, glaciers, the wind, and the sea. In

Part II we shall treat of those agencies whose sources are within the

earth, and among whose manifestations are volcanoes and earthquakes and the various movements of the earth's crust. As we study each agency we shall notice not only how it does its work, but also the records which it leaves in the rock structures and the land forms which it produces. With this preparation we shall be able in

Part III to read in the records of the rocks the

history of our planet and the successive forms of life which have dwelt upon it.

PART I

EXTERNAL GEOLOGICAL AGENCIES

CHAPTER I

THE WORK OF THE WEATHER

In our excursion to the valley with sandstone ledges we witnessed a process which is going forward in all lands. Everywhere the rocks are crumbling away; their fragments are creeping down hillsides to the stream ways and are carried by the streams to the sea, where they are rebuilt into rocky layers. When again the rocks are lifted to form land the process will begin anew; again they will crumble and creep down slopes and be washed by streams to the sea. Let us begin our study of this long cycle of change at the point where rocks disintegrate and decay under the action of the weather. In studying now a few outcrops and quarries we shall learn a little of some common rocks and how they weather away.

STRATIFICATION AND JOINTING. At the sandstone ledges we saw that the rock was divided into parallel layers. The thicker layers are known as STRATA, and the thin leaves into which each stratum may sometimes be split are termed LAMINAE. To a greater or less degree these layers differ from each other in fineness of grain, showing that the material has been sorted. The planes which divide them are called BEDDING PLANES.

Besides the bedding planes there are other division planes, which cut across the strata from top to bottom. These are found in all rocks and are known as joints. Two sets of joints, running at about right angles to each other, together with the bedding planes, divide the sandstone into quadrangular blocks.

SANDSTONE. Examining a piece of sandstone we find it composed of grains quite like those of river sand or of sea beaches. Most of the grains are of a clear glassy mineral called quartz. These quartz grains are very hard and will scratch the steel of a knife blade. They are not affected by acid, and their broken surfaces are irregular like those of broken glass.

The grains of sandstone are held together by some cement. This may be calcareous, consisting of soluble carbonate of lime. In brown sandstones the cement is commonly ferruginous,--hydrated iron oxide, or iron rust, forming the bond, somewhat as in the case of iron nails which have rusted together. The strongest and most lasting cement is siliceous, and sand rocks whose grains are closely cemented by silica, the chemical substance of which quartz is made, are known as quartzites.

We are now prepared to understand how sandstone is affected by the action of the weather. On ledges where the rock is exposed to view its surface is more or less discolored and the grains are loose and may be rubbed off with the finger. On gentle slopes the rock is covered with a soil composed of sand, which evidently is crumbled sandstone, and dark carbonaceous matter derived from the decay of vegetation. Clearly it is by the dissolving of the cement that the rock thus breaks down to loose sand. A piece of sandstone with calcareous cement, or a bit of old mortar, which is really an artificial stone also made of sand cemented by lime, may be treated in a test tube with hydrochloric acid to illustrate the process.

A LIMESTONE QUARRY. Here also we find the rock stratified and jointed (Fig. 2). On the quarry face the rock is distinctly seen to be altered for some distance from its upper surface. Below the altered zone the rock is sound and is quarried for building; but the altered upper layers are too soft and broken to be used for this purpose. If the limestone is laminated, the laminae here have split apart, although below they hold fast together. Near the surface the stone has become rotten and crumbles at the touch, while on the top it has completely broken down to a thin layer of limestone meal, on which rests a fine reddish clay.

Limestone is made of minute grains of carbonate of lime all firmly held together by a calcareous cement. A piece of the stone placed in a test tube with hydrochloric acid dissolves with brisk effervescence, leaving the insoluble impurities, which were disseminated through it, at the bottom of the tube as a little clay.

We can now understand the changes in the upper layers of the quarry. At the surface of the rock the limestone has completely dissolved, leaving the insoluble residue as a layer of reddish clay. Immediately below the clay the rock has disintegrated into meal where the cement between the limestone grains has been removed, while beneath this the laminae are split apart where the cement has been dissolved only along the planes of lamination where the stone is more porous. As these changes in the rock are greatest at the surface and diminish downward, we infer that they have been caused by agents working downward from the surface.

At certain points these agencies have been more effective than elsewhere. The upper rock surface is pitted. Joints are widened as they approach the surface, and along these seams we may find that the rock is altered even down to the quarry floor.

A SHALE PIT. Let us now visit some pit where shale--a laminated and somewhat hardened clay--is quarried for the manufacture of brick. The laminae of this fine-grained rock may be as thin as cardboard in places, and close joints may break the rock into small rhombic blocks. On the upper surface we note that the shale has weathered to a clayey soil in which all traces of structure have been destroyed. The clay and the upper layers of the shale beneath it are reddish or yellow, while in many cases the color of the unaltered rock beneath is blue.

THE SEDIMENTARY ROCKS. The three kinds of layered rocks whose acquaintance we have made--sandstone, limestone, and shale--are the leading types of the great group of stratified, or sedimentary, rocks. This group includes all rocks made of sediments, their materials having settled either in water upon the bottoms of rivers, lakes, or seas, or on dry land, as in the case of deposits made by the wind and by glaciers. Sedimentary rocks are divided into the fragmental rocks--which are made of fragments, either coarse or fine-- and the far less common rocks which are constituted of chemical precipitates.

The sedimentary rocks are divided according to their composition into the following classes:

1. The arenaceous, or quartz rocks, including beds of loose sand and gravel, sandstone, quartzite, and conglomerate (a rock made of cemented rounded gravel or pebbles).

2. The calcareous, or lime rocks, including limestone and a soft white rock formed of calcareous powder known as chalk.

3. The argillaceous, or clay rocks, including muds, clays, and shales. These three classes pass by mixture into one another. Thus there are limy and clayey sandstones, sandy and clayey limestones, and sandy and limy shales.

GRANITE. This familiar rock may be studied as an example of the second great group of rocks,--the unstratified, or igneous rocks. These are not made of cemented sedimentary grains, but of interlocking crystals which have crystallized from a molten mass. Examining a piece of granite, the most conspicuous crystals which meet the eye are those of feldspar. They are commonly pink, white, or yellow, and break along smooth cleavage planes which reflect the light like tiny panes of glass. Mica may be recognized by its glittering plates, which split into thin elastic scales. A third mineral, harder than steel, breaking along irregular surfaces like broken glass, we identify as quartz.

How granite alters under the action of the weather may be seen in outcrops where it forms the bed rock, or country rock, underlying the loose formations of the surface, and in many parts of the northern states where granite bowlders and pebbles more or less decayed may be found in a surface sheet of stony clay called the drift. Of the different minerals composing granite, quartz alone remains unaltered. Mica weathers to detached flakes which have lost their elasticity. The feldspar crystals have lost their luster and hardness, and even have decayed to clay. Where long- weathered granite forms the country rock, it often may be cut with spade or trowel for several feet from the surface, so rotten is the feldspar, and here the rock is seen to break down to a clayey soil containing grains of quartz and flakes of mica.

These are a few simple illustrations of the surface changes which some of the common kinds of rocks undergo. The agencies by which these changes are brought about we will now take up under two divisions,--CHEMICAL AGENCIES producing rock decay and MECHANICAL AGENCIES producing rock disintegration.

THE CHEMICAL WORK OF WATER

As water falls on the earth in rain it has already absorbed from the air carbon dioxide (carbonic acid gas) and oxygen. As it sinks into the ground and becomes what is termed ground water, it takes into solution from the soil humus acids and carbon dioxide, both of which are constantly being generated there by the decay of organic matter. So both rain and ground water are charged with active chemical agents, by the help of which they corrode and rust and decompose all rocks to a greater or less degree. We notice now three of the chief chemical processes concerned in weathering,-- solution, the formation of carbonates, and oxidation.

SOLUTION. Limestone, although so little affected by pure water that five thousand gallons would be needed to dissolve a single pound, is easily dissolved in water charged with carbon dioxide. In limestone regions well water is therefore "hard." On boiling the water for some time the carbon dioxide gas is expelled, the whole of the lime carbonate can no longer be held in solution, and much of it is thrown down to form a crust or "scale" in the kettle or in the tubes of the steam boiler. All waters which flow over limestone rocks or soak through them are constantly engaged in dissolving them away, and in the course of time destroy beds of vast extent and great thickness.

The upper surface of limestone rocks becomes deeply pitted, as we saw in the limestone quarry, and where the mantle of waste has been removed it may be found so intricately furrowed that it is difficult to traverse.

Beds of rock salt buried among the strata are dissolved by seeping water, which issues in salt springs. Gypsum, a mineral composed of hydrated sulphate of lime, and so soft that it may be scratched with the finger nail, is readily taken up by water, giving to the water of wells and springs a peculiar hardness difficult to remove.

The dissolving action of moisture may be noted on marble tombstones of some age, marble being a limestone altered by heat and pressure and composed of crystalline grains. By assuming that the date on each monument marks the year of its erection, one may estimate how many years on the average it has taken for weathering to loosen fine grains on the polished surface, so that they may be rubbed off with the finger, to destroy the polish, to round the sharp edges of tool marks in the lettering, and at last to open cracks and seams and break down the stone. We may notice also whether the gravestones weather more rapidly on the sunny or the shady side, and on the sides or on the top.

The weathered surface of granular limestone containing shells shows them standing in relief. As the shells are made of crystalline carbonate of lime, we may infer whether the carbonate of lime is less soluble in its granular or in its crystalline condition.

THE FORMATION OF CARBONATES. In attacking minerals water does more than merely take them into solution. It decomposes them, forming new chemical compounds of which the carbonates are among the most important. Thus feldspar consists of the insoluble silicate of alumina, together with certain alkaline silicates which are broken up by the action of water containing carbon dioxide, forming alkaline carbonates. These carbonates are freely soluble and contribute potash and soda to soils and river waters. By the removal of the soluble ingredients of feldspar there is left the silicate of alumina, united with water or hydrated, in the condition of a fine plastic clay which, when white and pure, is known as KAOLIN and is used in the manufacture of porcelain. Feldspathic rocks which contain no iron compounds thus weather to whitish crusts, and even apparently sound crystals of feldspar, when ground to thin slices and placed under the microscope, may be seen to be milky in color throughout because an internal change to kaolin has begun.

OXIDATION. Rocks containing compounds of iron weather to reddish crusts, and the seams of these rocks are often lined with rusty films. Oxygen and water have here united with the iron, forming hydrated iron oxide. The effects of oxidation may be seen in the alteration of many kinds of rocks and in red and yellow colors of soils and subsoils.

Pyrite is a very hard mineral of a pale brass color, found in scattered crystals in many rocks, and is composed of iron and sulphur (iron sulphide). Under the attack of the weather it takes up oxygen, forming iron sulphate (green vitriol), a soluble compound, and insoluble hydrated iron oxide, which as a mineral is known as limonite. Several large masses of iron sulphide were placed some years ago on the lawn in front of the National Museum at Washington. The mineral changed so rapidly to green vitriol that enough of this poisonous compound was washed into the ground to kill the roots of the surrounding grass.

AGENTS OF MECHANICAL DISINTEGRATION

HEAT AND COLD. Rocks exposed to the direct rays of the sun become strongly heated by day and expand. After sunset they rapidly cool and contract. When the difference in temperature between day and night is considerable, the repeated strains of sudden expansion and contraction at last become greater than the rocks can bear, and they break, for the same reason that a glass cracks when plunged into boiling water (Fig. 5).

Rocks are poor conductors of heat, and hence their surfaces may become painfully hot under the full blaze of the sun, while the interior remains comparatively cool. By day the surface shell expands and tends to break loose from the mass of the stone. In cooling in the evening the surface shell suddenly contracts on the unyielding interior and in time is forced off in scales.

Many rocks, such as granite, are made up of grains of various minerals which differ in color and in their capacity to absorb heat, and which therefore contract and expand in different ratios. In heating and cooling these grains crowd against their neighbors and tear loose from them, so that finally the rock disintegrates into sand.

The conditions for the destructive action of heat and cold are most fully met in arid regions when vegetation is wanting for lack of sufficient rain. The soil not being held together by the roots of plants is blown away over large areas, leaving the rocks bare to the blazing sun in a cloudless sky. The air is dry, and the heat received by the earth by day is therefore rapidly radiated at night into space. There is a sharp and sudden fall of temperature after sunset, and the rocks, strongly heated by day, are now chilled perhaps even to the freezing point.

In the Sahara the thermometer has been known to fall 131 degrees F. within a few hours. In the light air of the Pamir plateau in central Asia a rise of 90 degrees F. has been recorded from seven o'clock in the morning to one o'clock in the afternoon. On the mountains of southwestern Texas there are frequently heard crackling noises as the rocks of that arid region throw off scales from a fraction of an inch to four inches in thickness, and loud reports are made as huge bowlders split apart. Desert pebbles weakened by long exposure to heat and cold have been shivered to fine sharp-pointed fragments on being placed in sand heated to 180 degrees F. Beds half a foot thick, forming the floor of limestone quarries in Wisconsin, have been known to buckle and arch and break to fragments under the heat of the summer sun.

FROST. By this term is meant the freezing and thawing of water contained in the pores and crevices of rocks. All rocks are more or less porous and all contain more or less water in their pores. Workers in stone call this "quarry water," and speak of a stone as "green" before the quarry water has dried out. Water also seeps along joints and bedding planes and gathers in all seams and crevices. Water expands in freezing, ten cubic inches of water freezing to about eleven cubic inches of ice. As water freezes in the rifts and pores of rocks it expands with the irresistible force illustrated in the freezing and breaking of water pipes in winter. The first rift in the rock, perhaps too narrow to be seen, is widened little by little by the wedges of successive frosts, and finally the rock is broken into detached blocks, and these into angular chip-stone by the same process.

It is on mountain tops and in high latitudes that the effects of frost are most plainly seen. "Every summit" says Whymper, "amongst the rock summits upon which I have stood has been nothing but a piled-up heap of fragments" (Fig. 7). In Iceland, in Spitsbergen, in Kamchatka, and in other frigid lands large areas are thickly strewn with sharp-edged fragments into which the rock has been shattered by frost.

ORGANIC AGENTS

We must reckon the roots of plants and trees among the agents which break rocks into pieces. The tiny rootlet in its search for food and moisture inserts itself into some minute rift, and as it grows slowly wedges the rock apart. Moreover, the acids of the root corrode the rocks with which they are in contact. One may sometimes find in the soil a block of limestone wrapped in a mesh of roots, each of which lies in a little furrow where it has eaten into the stone.

Rootless plants called lichens often cover and corrode rocks as yet bare of soil; but where lichens are destroying the rock less rapidly than does the weather, they serve in a way as a protection.

CONDITIONS FAVORING DISINTEGRATION AND DECAY. The disintegration of rocks under frost and temperature changes goes on most rapidly in cold and arid climates, and where vegetation is scant or absent. On the contrary, the decay of rocks under the chemical action of water is favored by a warm, moist climate and abundant vegetation. Frost and heat and cold can only act within the few feet from the surface to which the necessary temperature changes are limited, while water penetrates and alters the rocks to great depths.

The pupil may explain.

In what ways the presence of joints and bedding planes assists in the breaking up and decay of rocks under the action of the weather.

Why it is a good rule of stone masons never to lay stones on edge, but always on their natural bedding planes.

Why stones fresh from the quarry sometimes go to pieces in early winter, when stones which have been quarried for some months remain uninjured.

Why quarrymen in the northern states often keep their quarry floors flooded during winter.

Why laminated limestone should not be used for curbstone.

Why rocks composed of layers differing in fineness of grain and in ratios of expansion do not make good building stone.

Fine-grained rocks with pores so small that capillary attraction keeps the water which they contain from readily draining away are more apt to hold their pores ten elevenths full of water than are rocks whose pores are larger. Which, therefore, are more likely to be injured by frost?

Which is subject to greater temperature changes, a dark rock or one of a light color? the north side or the south side of a valley?

THE MANTLE OF ROCK WASTE

We have seen that rocks are everywhere slowly wasting away. They are broken in pieces by frost, by tree roots, and by heat and cold. They dissolve and decompose under the chemical action of water and the various corrosive substances which it contains, leaving their insoluble residues as residual clays and sands upon the surface. As a result there is everywhere forming a mantle of rock waste which covers the land. It is well to imagine how the country would appear were this mantle with its soil and vegetation all scraped away or had it never been formed. The surface of the land would then be everywhere of bare rock as unbroken as a quarry floor.

THE THICKNESS OF THE MANTLE. In any locality the thickness of the mantle of rock waste depends as much on the rate at which it is constantly being removed as on the rate at which it is forming. On the face of cliffs it is absent, for here waste is removed as fast as it is made. Where waste is carried away more slowly than it is produced, it accumulates in time to great depth.

The granite of Pikes Peak is disintegrated to a depth of twenty feet. In the city of Washington granite rock is so softened to a depth of eighty feet that it can be removed with pick and shovel. About Atlanta, Georgia, the rocks are completely rotted for one hundred feet from the surface, while the beginnings of decay may be noticed at thrice that depth. In places in southern Brazil the rock is decomposed to a depth of four hundred feet.

In southwestern Wisconsin a reddish residual clay has an average depth of thirteen feet on broad uplands, where it has been removed to the least extent. The country rock on which it rests is a limestone with about ten per cent of insoluble impurities. At least how thick, then, was that portion of the limestone which has rotted down to the clay?

DISTINGUISHING CHARACTERISTICS OF RESIDUAL WASTE. We must learn to distinguish waste formed in place by the action of the weather from the products of other geological agencies. Residual waste is unstratified. It contains no substances which have not been derived from the weathering of the parent rock. There is a gradual transition from residual waste into the unweathered rock beneath. Waste resting on sound rock evidently has been shifted and was not formed in place.

In certain regions of southern Missouri the land is covered with a layer of broken flints and red clay, while the country rock is limestone. The limestone contains nodules of flint, and we may infer that it has been by the decay and removal of thick masses of limestone that the residual layer of clay and flints has been left upon the surface. Flint is a form of quartz, dull-lustered, usually gray or blackish in color, and opaque except on thinnest edges, where it is translucent.

Over much of the northern states there is spread an unstratified stony clay called the drift. It often rests on sound rocks. It contains grains of sand, pebbles, and bowlders composed of many different minerals and rocks that the country rock cannot furnish. Hence the drift cannot have been formed by the decay of the rock of the region. A shale or limestone, for example, cannot waste to a clay containing granite pebbles. The origin of the drift will be explained in subsequent chapters.

The differences in rocks are due more to their soluble than to their insoluble constituents. The latter are few in number and are much the same in rocks of widely different nature, being chiefly quartz, silicate of alumina, and iron oxide. By the removal of their soluble parts very many and widely different rocks rot down to a residual clay gritty with particles of quartz and colored red or yellow with iron oxide.

In a broad way the changes which rocks undergo in weathering are an adaptation to the environment in which they find themselves at the earth's surface,--an environment different from that in which they were formed under sea or under ground. In open air, where they are attacked by various destructive agents, few of the rock- making minerals are stable compounds except quartz, the iron oxides, and the silicate of alumina; and so it is to one or more of these comparatively insoluble substances that most rocks are reduced by long decay.

Which produces a mantle of finer waste, frost or chemical decay? which a thicker mantle? In what respects would you expect that the mantle of waste would differ in warm humid lands like India, in frozen countries like Alaska, and in deserts such as the Sahara?

THE SOIL. The same agencies which produce the mantle of waste are continually at work upon it, breaking it up into finer and finer particles and causing its more complete decay. Thus on the surface, where the waste has weathered longest, it is gradually made fine enough to support the growth of plants, and is then known as soil. The coarser waste beneath is sometimes spoken of as subsoil. Soil usually contains more or less dark, carbonaceous, decaying organic matter, called humus, and is then often termed the humus layer. Soil forms not only on waste produced in place from the rock beneath, but also on materials which have been transported, such as sheets of glacial drift and river deposits. Until rocks are reduced to residual clays the work of the weather is more rapid and effective on the fragments of the mantle of waste than on the rocks from which waste is being formed. Why?

Any fresh excavation of cellar or cistern, or cut for road or railway, will show the characteristics of the humus layer. It may form only a gray film on the surface, or we may find it a layer a foot or more thick, dark, or even black, above, and growing gradually lighter in color as it passes by insensible gradations into the subsoil. In some way the decaying vegetable matter continually forming on the surface has become mingled with the material beneath it.

HOW HUMUS AND THE SUBSOIL ARE MINGLED. The mingling of humus and the subsoil is brought about by several means. The roots of plants penetrate the waste, and when they die leave their decaying substance to fertilize it. Leaves and stems falling on the surface are turned under by several agents. Earthworms and other animals whose home is in the waste drag them into their burrows either for food or to line their nests. Trees overthrown by the wind, roots and all, turn over the soil and subsoil and mingle them together. Bacteria also work in the waste and contribute to its enrichment. The animals living in the mantle do much in other ways toward the making of soil. They bring the coarser fragments from beneath to the surface, where the waste weathers more rapidly. Their burrows allow air and water to penetrate the waste more freely and to affect it to greater depths.

ANTS. In the tropics the mantle of waste is worked over chiefly by ants. They excavate underground galleries and chambers, extending sometimes as much as fourteen feet below the surface, and build mounds which may reach as high above it. In some parts of Paraguay and southern Brazil these mounds, like gigantic potato hills, cover tracts of considerable area.

In search for its food--the dead wood of trees--the so-called white ant constructs runways of earth about the size of gas pipes, reaching from the base of the tree to the topmost branches. On the plateaus of central Africa explorers have walked for miles through forests every tree of which was plastered with these galleries of mud. Each grain of earth used in their construction is moistened and cemented by slime as it is laid in place by the ant, and is thus acted on by organic chemical agents. Sooner or later these galleries are beaten down by heavy rains, and their fertilizing substances are scattered widely by the winds.

EARTHWORMS. In temperate regions the waste is worked over largely by earthworms. In making their burrows worms swallow earth in order to extract from it any nutritive organic matter which it may contain. They treat it with their digestive acids, grind it in their stony gizzards, and void it in castings on the surface of the ground. It was estimated by Darwin that in many parts of England each year, on every acre, more than ten tons of earth pass through the bodies of earthworms and are brought to the surface, and that every few years the entire soil layer is thus worked over by them.

In all these ways the waste is made fine and stirred and enriched. Grain by grain the subsoil with its fresh mineral ingredients is brought to the surface, and the rich organic matter which plants and animals have taken from the atmosphere is plowed under. Thus Nature plows and harrows on "the great world's farm" to make ready and ever to renew a soil fit for the endless succession of her crops.

The world processes by which rocks are continually wasting away are thus indispensable to the life of plants and animals. The organic world is built on the ruins of the inorganic, and because the solid rocks have been broken down into soil men are able to live upon the earth.

SOLAR ENERGY. The source of the energy which accomplishes all this necessary work is the sun. It is the radiant energy of the sun which causes the disintegration of rocks, which lifts vapor into the atmosphere to fall as rain, which gives life to plants and animals. Considering the earth in a broad way, we may view it as a globe of solid rock,--the lithosphere,--surrounded by two mobile envelopes: the envelope of air,--THE ATMOSPHERE, and the envelope of water,--THE HYDROSPHERE. Under the action of solar energy these envelopes are in constant motion. Water from the hydrosphere is continually rising in vapor into the atmosphere, the air of the atmosphere penetrates the hydrosphere,--for its gases are dissolved in all waters,--and both air and water enter and work upon the solid earth. By their action upon the lithosphere they have produced a third envelope,--the mantle of rock waste.

This envelope also is in movement, not indeed as a whole, but particle by particle. The causes which set its particles in motion, and the different forms which the mantle comes to assume, we will now proceed to study.

MOVEMENTS OF THE MANTLE OF ROCK WASTE

At the sandstone ledges which we first visited we saw not only that the rocks were crumbling away, but also that grains and fragments of them were creeping down the slopes of the valley to the stream and were carried by it onward toward the sea. This process is going on everywhere. Slowly it may be, and with many interruptions, but surely, the waste of the land moves downward to the sea. We may divide its course into two parts,--the path to the stream, which we will now consider, and its carriage onward by the stream, which we will defer to a later chapter.

GRAVITY. The chief agent concerned in the movement of waste is gravity. Each particle of waste feels the unceasing downward pull of the earth's mass and follows it when free to do so. All agencies which produce waste tend to set its particles free and in motion, and therefore cooperate with gravity. On cliffs, rocks fall when wedged off by frost or by roots of trees, and when detached by any other agency. On slopes of waste, water freezes in chinks between stones, and in pores between particles of soil, and wedges them apart. Animals and plants stir the waste, heat expands it, cold contracts it, the strokes of the raindrops drive loose particles down the slope and the wind lifts and lets them fall. Of all these movements, gravity assists those which are downhill and retards those which are uphill. On the whole, therefore, the downhill movements prevail, and the mantle of waste, block by block and grain by grain, creeps along the downhill path.

A slab of sandstone laid on another of the same kind at an angle of 17 degrees and left in the open air was found to creep down the slope at the rate of a little more than a millimeter a month. Explain why it did so.

RAIN. The most efficient agent in the carriage of waste to the streams is the rain. It moves particles of soil by the force of the blows of the falling drops, and washes them down all slopes to within reach of permanent streams. On surfaces unprotected by vegetation, as on plowed fields and in arid regions, the rain wears furrows and gullies both in the mantle of waste and in exposures of unaltered rock (Fig. 17).

At the foot of a hill we may find that the soil has accumulated by creep and wash to the depth of several feet; while where the hillside is steepest the soil may be exceedingly thin, or quite absent, because removed about as fast as formed. Against the walls of an abbey built on a slope in Wales seven hundred years ago, the creeping waste has gathered on the uphill side to a depth of seven feet. The slow-flowing sheet of waste is often dammed by fences and walls, whose uphill side gathers waste in a few years so as to show a distinctly higher surface than the downhill side, especially in plowed fields where the movement is least checked by vegetation.

TALUS. At the foot of cliffs there is usually to be found a slope of rock fragments which clearly have fallen from above. Such a heap of waste is known as talus. The amount of talus in any place depends both on the rate of its formation and the rate of its removal. Talus forms rapidly in climates where mechanical disintegration is most effective, where rocks are readily broken into blocks because closely jointed and thinly bedded rather than massive, and where they are firm enough to be detached in fragments of some size instead of in fine grains. Talus is removed slowly where it decays slowly, either because of the climate or the resistance of the rock. It may be rapidly removed by a stream flowing along its base.

In a moist climate a soluble rock, such as massive limestone, may form talus little if any faster than the talus weathers away. A loose-textured sandstone breaks down into incoherent sand grains, which in dry climates, where unprotected by vegetation, may be blown away as fast as they fall, leaving the cliff bare to the base. Cliffs of such slow-decaying rocks as quartzite and granite when closely jointed accumulate talus in large amounts.

Talus slopes may be so steep as to reach THE ANGLE OF REPOSE, i.e. the steepest angle at which the material will lie. This angle varies with different materials, being greater with coarse and angular fragments than with fine rounded grains. Sooner or later a talus reaches that equilibrium where the amount removed from its surface just equals that supplied from the cliff above. As the talus is removed and weathers away its slope retreats together with the retreat of the cliff, as seen in Figure 9.

GRADED SLOPES. Where rocks weather faster than their waste is carried away, the waste comes at last to cover all rocky ledges. On the steeper slopes it is coarser and in more rapid movement than on slopes more gentle, but mountain sides and hills and plains alike come to be mantled with sheets of waste which everywhere is creeping toward the streams. Such unbroken slopes, worn or built to the least inclination at which the waste supplied by weathering can be urged onward, are known as GRADED SLOPES.

Of far less importance than the silent, gradual creep of waste, which is going on at all times everywhere about us, are the startling local and spasmodic movements which we are now to describe.

AVALANCHES. On steep mountain sides the accumulated snows of winter often slip and slide in avalanches to the valleys below. These rushing torrents of snow sweep their tracks clean of waste and are one of Nature's normal methods of moving it along the downhill path.

LANDSLIDES. Another common and abrupt method of delivering waste to streams is by slips of the waste mantle in large masses. After long rains and after winter frosts the cohesion between the waste and the sound rock beneath is loosened by seeping water underground. The waste slips on the rock surface thus lubricated and plunges down the mountain side in a swift roaring torrent of mud and stones.

We may conveniently mention here a second type of landslide, where masses of solid rock as well as the mantle of waste are involved in the sudden movement. Such slips occur when valleys have been rapidly deepened by streams or glaciers and their sides have not yet been graded. A favorable condition is where the strata dip (i.e. incline downwards) towards the valley (Fig. 11), or are broken by joint planes dipping in the same direction. The upper layers, including perhaps the entire mountain side, have been cut across by the valley trench and are left supported only on the inclined surface of the underlying rocks. Water may percolate underground along this surface and loosen the cohesion between the upper and the underlying strata by converting the upper surface of a shale to soft wet clay, by dissolving layers of a limestone, or by removing the cement of a sandstone and converting it into loose sand. When the inclined surface is thus lubricated the overlying masses may be launched into the valley below. The solid rocks are broken and crushed in sliding and converted into waste consisting, like that of talus, of angular unsorted fragments, blocks of all sizes being mingled pellmell with rock meal and dust. The principal effects of landslides may be gathered from the following examples.

At Gohna, India, in 1893, the face of a spur four thousand feet high, of the lower ranges of the Himalayas, slipped into the gorge of the headwaters of the Ganges River in successive rock falls which lasted for three days. Blocks of stone were projected for a mile, and clouds of limestone dust were spread over the surrounding country. The debris formed a dam one thousand feet high, extending for two miles along the valley. A lake gathered behind this barrier, gradually rising until it overtopped it in a little less than a year. The upper portion of the dam then broke, and a terrific rush of water swept down the valley in a wave which, twenty miles away, rose one hundred and sixty feet in height. A narrow lake is still held by the strong base of the dam.

In 1896, after forty days of incessant rain, a cliff of sandstone slipped into the Yangtse River in China, reducing the width of the channel to eighty yards and causing formidable rapids.

At Flims, in Switzerland, a prehistoric landslip flung a dam eighteen hundred feet high across the headwaters of the Rhine. If spread evenly over a surface of twenty-eight square miles, the material would cover it to a depth of six hundred and sixty feet. The barrier is not yet entirely cut away, and several lakes are held in shallow basins on its hummocky surface.

A slide from the precipitous river front of the citadel hill of Quebec, in 1889, dashed across Champlain Street, wrecking a number of houses and causing the death of forty-five persons. The strata here are composed of steeply dipping slate.

In lofty mountain ranges there may not be a single valley without its traces of landslides, so common there is this method of the movement of waste, and of building to grade over-steepened slopes.

ROCK SCULPTURE BY WEATHERING

We are now to consider a few of the forms into which rock masses are carved by the weather.

BOWLDERS OF WEATHERING. In many quarries and outcrops we may see that the blocks into which one or more of the uppermost layers have been broken along their joints and bedding planes are no longer angular, as are those of the layers below. The edges and corners of these blocks have been worn away by the weather. Such rounded cores, known as bowlders of weathering, are often left to strew the surface.

DIFFERENTIAL WEATHERING. This term covers all cases in which a rock mass weathers differently in different portions. Any weaker spots or layers are etched out on the surface, leaving the more resistant in relief. Thus massive limestones become pitted where the weather drills out the weaker portions. In these pits, when once they are formed, moisture gathers, a little soil collects, vegetation takes root, and thus they are further enlarged until the limestone may be deeply honeycombed.

On the sides of canyons, and elsewhere where the edges of strata are exposed, the harder layers project as cliffs, while the softer weather back to slopes covered with the talus of the harder layers above them. It is convenient to call the former cliff makers and the latter slope makers.

Differential weathering plays a large part in the sculpture of the land. Areas of weak rock are wasted to plains, while areas of hard rock adjacent are still left as hills and mountain ridges, as in the valleys and mountains of eastern Pennsylvania. But in such instances the lowering of the surface of the weaker rock is also due to the wear of streams, and especially to the removal by them from the land of the waste which covers and protects the rocks beneath.

Rocks owe their weakness to several different causes. Some, such as beds of loose sand, are soft and easily worn by rains; some, as limestone and gypsum for example, are soluble. Even hard insoluble rocks are weak under the attack of the weather when they are closely divided by joints and bedding planes and are thus readily broken up into blocks by mechanical agencies.

OUTLIERS AND MONUMENTS. As cliffs retreat under the attack of the weather, portions are left behind where the rock is more resistant or where the attack for any reason is less severe. Such remnant masses, if large, are known as outliers. When

Note the rain furrows on the slope at the foot of the monuments. In the foreground are seen fragments of petrified trunks of trees, composed of silica and extremely resistant to the weather. On the removal of the rock layers in which these fragments were imbedded they are left to strew the surface in the same way as are the residual flints of southern Missouri. flat-topped, because of the protection of a resistant horizontal capping layer, they are termed mesas,--a term applied also to the flat-topped portions of dissected plateaus (Fig. 129). Retreating cliffs may fall back a number of miles behind their outliers before the latter are finally consumed.

Monuments are smaller masses and may be but partially detached from the cliff face. In the breaking down of sheets of horizontal strata, outliers grow smaller and smaller and are reduced to massive rectangular monuments resembling castles (Fig. 17). The rock castle falls into ruin, leaving here and there an isolated tower; the tower crumbles to a lonely pillar, soon to be overthrown. The various and often picturesque shapes of monuments depend on the kind of rock, the attitude of the strata, and the agent by which they are chiefly carved. Thus pillars may have a capital formed of a resistant stratum. Monuments may be undercut and come to rest on narrow pedestals, wherever they weather more rapidly near the ground, either because of the greater moisture there, or--in arid climates--because worn at their base by drifting sands.

Stony clays disintegrating under the rain often contain bowlders which protect the softer material beneath from the vertical blows of raindrops, and thus come to stand on pedestals of some height. One may sometimes see on the ground beneath dripping eaves pebbles left in the same way, protecting tiny pedestals of sand.

MOUNTAIN PEAKS AND RIDGES. Most mountains have been carved out of great broadly uplifted folds and blocks of the earth's crust. Running water and glacier ice have cut these folds and blocks into masses divided by deep valleys; but it is by the weather, for the most part, that the masses thus separated have been sculptured to the present forms of the individual peaks and ridges.

Frost and heat and cold sculpture high mountains to sharp, tusklike peaks and ragged, serrate crests, where their waste is readily removed.

The Matterhorn of the Alps is a famous example of a mountain peak whose carving by the frost and other agents is in active progress. On its face "scarcely a rock anywhere is firmly attached," and the fall of loosened stones is incessant. Mountain climbers who have camped at its base tell how huge rocks from time to time come leaping down its precipices, followed by trains of dislodged smaller fragments and rock dust; and how at night one may trace the course of the bowlders by the sparks which they strike from the mountain walls. Mount Assiniboine, Canada (Fig. 20), resembles the Matterhorn in form and has been carved by the same agencies.

"The Needles" of Arizona are examples of sharp mountain peaks in a warm arid region sculptured chiefly by temperature changes.

Chemical decay, especially when carried on beneath a cover of waste and vegetation, favors the production of rounded knobs and dome-shaped mountains.

THE WEATHER CURVE. We have seen that weathering reduces the angular block quarried by the frost to a rounded bowlder by chipping off its corners and smoothing away its edges. In much the same way weathering at last reduces to rounded hills the earth blocks cut by streams or formed in any other way. High mountains may at first be sculptured by the weather to savage peaks (Fig. 181), but toward the end of their life history they wear down to rounded hills (Fig. 182). The weather curve, which may be seen on the summits of low hills (Fig. 21), is convex upward.

In Figure 22, representing a cubic block of stone whose faces are a yard square, how many square feet of surface are exposed to the weather by a cubic foot at a corner a; by one situated in the middle of an edge b; by one in the center of a side c? How much faster will a and b weather than c, and what will be the effect on the shape of the block?

THE COOPERATION OF VARIOUS AGENCIES IN ROCK SCULPTURE. For the sake of clearness it is necessary to describe the work of each geological agent separately. We must not forget, however, that in Nature no agent works independently and alone; that every result is the outcome of a long chain of causes. Thus, in order that the mountain peak may be carved by the agents of disintegration, the waste must be rapidly removed,--a work done by many agents, including some which we are yet to study; and in order that the waste may be removed as fast as formed, the region must first have been raised well above the level of the sea, so that the agents of transportation could do their work effectively. The sculpture of the rocks is accomplished only by the cooperation of many forces.

The constant removal of waste from the surface by creep and wash and carriage by streams is of the highest importance, because it allows the destruction of the land by means of weathering to go on as long as any land remains above sea level. If waste were not removed, it would grow to be so thick as to protect the rock beneath from further weathering, and the processes of destruction which we have studied would be brought to an end. The very presence of the mantle of waste over the land proves that on the whole rocks weather more rapidly than their waste is removed. The destruction of the land is going on as fast as the waste can be carried away.

We have now learned to see in the mantle of waste the record of the destructive action of the agencies of weathering on the rocks of the land surface. Similar records we shall find buried deeply among the rocks of the crust in old soils and in rocks pitted and decayed, telling of old land surfaces long wasted by the weather. Ever since the dry land appeared these agencies have been as now quietly and unceasingly at work upon it, and have ever been the chief means of the destruction of its rocks. The vast bulk of the stratified rocks of the earth's crust is made up almost wholly of the waste thus worn from ancient lands.

In studying the various geological agencies we must remember the almost inconceivable times in which they work. The slowest process when multiplied by the immense time in which it is carried on produces great results. The geologist looks upon the land forms of the earth's surface as monuments which record the slow action of weathering and other agents during the ages of the past. The mountain peak, the rounded hill, the wide plain which lies where hills and mountains once stood, tell clearly of the great results which slow processes will reach when given long time in which to do their work. We should accustom ourselves also to think of the results which weathering will sooner or later bring to pass. The tombstone and the bowlder of the field, which each year lose from their surfaces a few crystalline grains, must in time be wholly destroyed. The hill whose rocks are slowly rotting underneath a cover of waste must become lower and lower as the centuries and millenniums come and go, and will finally disappear. Even the mountains are crumbling away continually, and therefore are but fleeting features of the landscape.

CHAPTER II

THE WORK OF GROUND WATER

LAND WATERS. We have seen how large is the part that water plays at and near the surface of the land in the processes of weathering and in the slow movement of waste down all slopes to the stream ways. We now take up the work of water as it descends beneath the ground,--a corrosive agent still, and carrying in solution as its load the invisible waste of rocks derived from their soluble parts.

Land waters have their immediate source in the rainfall. By the heat of the sun water is evaporated from the reservoir of the ocean and from moist surfaces everywhere. Mingled as vapor with the air, it is carried by the winds over sea and land, and condensed it returns to the earth as rain or snow. That part of the rainfall which descends on the ocean does not concern us, but that which falls on the land accomplishes, as it returns to the sea, the most important work of all surface geological agencies.

The rainfall may be divided into three parts: the first DRIES UP, being discharged into the air by evaporation either directly from the soil or through vegetation; the second RUNS OFF over the surface to flood the streams; the third SOAKS IN the ground and is henceforth known as GROUND or UNDERGROUND WATER.

THE DESCENT OF GROUND WATER. Seeping through the mantle of waste, ground water soaks into the pores and crevices of the underlying rock. All rocks of the upper crust of the earth are more or less porous, and all drink in water. IMPERVIOUS ROCKS, such as granite, clay, and shale, have pores so minute that the water which they take in is held fast within them by capillary attraction, and none drains through. PERVIOUS ROCKS, on the other hand, such as many sandstones, have pore spaces so large that water filters through them more or less freely. Besides its seepage through the pores of pervious rocks, water passes to lower levels through the joints and cracks by which all rocks, near the surface are broken.

Even the closest-grained granite has a pore space of 1 in 400, while sandstone may have a pore space of 1 in 4. Sand is so porous that it may absorb a third of its volume of water, and a loose loam even as much as one half.

THE GROUND-WATER SURFACE is the name given the upper surface of ground water, the level below which all rocks are saturated. In dry seasons the ground-water surface sinks. For ground water is constantly seeping downward under gravity, it is evaporated in the waste and its moisture is carried upward by capillarity and the roots of plants to the surface to be evaporated in the air. In wet seasons these constant losses are more than made good by fresh supplies from that part of the rainfall which soaks into the ground, and the ground-water surface rises.

In moist climates the ground-water surface (Fig. 24) lies, as a rule, within a few feet of the land surface and conforms to it in a general way, although with slopes of less inclination than those of the hills and valleys. In dry climates permanent ground water may be found only at depths of hundreds of feet. Ground water is held at its height by the fact that its circulation is constantly impeded by capillarity and friction. If it were as free to drain away as are surface streams, it would sink soon after a rain to the level of the deepest valleys of the region.

WELLS AND SPRINGS. Excavations made in permeable rocks below the ground-water surface fill to its level and are known as wells. Where valleys cut this surface permanent streams are formed, the water either oozing forth along ill-defined areas or issuing at definite points called springs, where it is concentrated by the structure of the rocks. A level tract where the ground-water surface coincides with the surface of the ground is a swamp or marsh.

By studying a spring one may learn much of the ways and work of ground water. Spring water differs from that of the stream into which it flows in several respects. If we test the spring with a thermometer during successive months, we shall find that its temperature remains much the same the year round. In summer it is markedly cooler than the stream; in winter it is warmer and remains unfrozen while the latter perhaps is locked in ice. This means that its underground path must lie at such a distance from the surface that it is little affected by summer's heat and winter's cold.

While the stream is often turbid with surface waste washed into it by rains, the spring remains clear; its water has been filtered during its slow movement through many small underground passages and the pores of rocks. Commonly the spring differs from the stream in that it carries a far larger load of dissolved rock. Chemical analysis proves that streams contain various minerals in solution, but these are usually in quantities so small that they are not perceptible to the taste or feel. But the water of springs is often well charged with soluble minerals; in its slow, long journey underground it has searched out the soluble parts of the rocks through which it seeps and has dissolved as much of them as it could. When spring water is boiled away, the invisible load which it has carried is left behind, and in composition is found to be practically identical with that of the soluble ingredients of the country rock. Although to some extent the soluble waste of rocks is washed down surface slopes by the rain, by far the larger part is carried downward by ground water and is delivered to streams by springs.

In limestone regions springs are charged with calcium carbonate (the carbonate of lime), and where the limestone is magnesian they contain magnesium carbonate also. Such waters are "hard"; when used in washing, the minerals which they contain combine with the fatty acids of soap to form insoluble curdy compounds. When springs rise from rocks containing gypsum they are hard with calcium sulphate. In granite regions they contain more or less soda and potash from the decay of feldspar.

The flow of springs varies much less during the different seasons of the year than does that of surface streams. So slow is the movement of ground water through the rocks that even during long droughts large amounts remain stored above the levels of surface drainage.

MOVEMENTS OF GROUND WATER. Ground water is in constant movement toward its outlets. Its rate varies according to many conditions, but always is extremely slow. Even through loose sands beneath the beds of rivers it sometimes does not exceed a fifth of a mile a year.

In any region two zones of flow may be distinguished. The UPPER ZONE OF FLOW extends from the ground-water surface downward through the waste mantle and any permeable rocks on which the mantle rests, as far as the first impermeable layer, where the descending movement of the water is stopped. The DEEP ZONES OF FLOW occupy any pervious rocks which may be found below the impervious layer which lies nearest to the surface. The upper zone is a vast sheet of water saturating the soil and rocks and slowly seeping downward through their pores and interstices along the slopes to the valleys, where in part it discharges in springs and often unites also in a wide underflowing stream which supports and feeds the river (Fig. 24).

A city in a region of copious rains, built on the narrow flood plain of a river, overlooked by hills, depends for its water supply on driven wells, within the city limits, sunk in the sand a few yards from the edge of the stream. Are these wells fed by water from the river percolating through the sand, or by ground water on its way to the stream and possibly contaminated with the sewage of the town?

At what height does underground water stand in the wells of your region? Does it vary with the season? Have you ever known wells to go dry? It may be possible to get data from different wells and to draw a diagram showing the ground-water surface as compared with the surface of the ground.

FISSURE SPRINGS AND ARTESIAN WELLS. The DEEPER ZONES OF FLOW lie in pervious strata which are overlain by some impervious stratum. Such layers are often carried by their dip to great depths, and water may circulate in them to far below the level of the surface streams and even of the sea. When a fissure crosses a water- bearing stratum, or AQUIFIER, water is forced upward by the pressure of the weight of the water contained in the higher parts of the stratum, and may reach the surface as a fissure spring. A boring which taps such an aquifer is known as an artesian well, a name derived from a province in France where wells of this kind have been long in use. The rise of the water in artesian wells, and in fissure springs also, depends on the following conditions illustrated in Figure 29. The aquifer dips toward the region of the wells from higher ground, where it outcrops and receives its water. It is inclosed between an impervious layer above and water- tight or water-logged layers beneath. The weight of the column of water thus inclosed in the aquifer causes water to rise in the well, precisely as the weight of the water in a standpipe forces it in connected pipes to the upper stories of buildings.

Which will supply the larger region with artesian wells, an aquifer whose dip is steep or one whose dip is gentle? Which of the two aquifers, their thickness being equal, will have the larger outcrop and therefore be able to draw upon the larger amount of water from the rainfall? Illustrate with diagrams.

THE ZONE OF SOLUTION. Near the surface, where the circulation of ground water is most active, it oxidizes, corrodes, and dissolves the rocks through which it passes. It leaches soils and subsoils of their lime and other soluble minerals upon which plants depend for their food. It takes away the soluble cements of rocks; it widens fissures and joints and opens winding passages along the bedding planes; it may even remove whole beds of soluble rocks, such as rock salt, limestone, or gypsum. The work of ground water in producing landslides has already been noticed. The zone in which the work of ground water is thus for the most part destructive we may call the zone of solution.

CAVES. In massive limestone rocks, ground water dissolves channels which sometimes form large caves (Fig. 30). The necessary conditions for the excavation of caves of great size are well shown in central Kentucky, where an upland is built throughout of thick horizontal beds of limestone. The absence of layers of insoluble or impervious rock in its structure allows a free circulation of ground water within it by the way of all natural openings in the rock. These water ways have been gradually enlarged by solution and wear until the upland is honeycombed with caves. Five hundred open caverns are known in one county.

Mammoth Cave, the largest of these caverns, consists of a labyrinth of chambers and winding galleries whose total length is said to be as much as thirty miles. One passage four miles long has an average width of about sixty feet and an average height of forty feet. One of the great halls is three hundred feet in width and is overhung by a solid arch of limestone one hundred feet above the floor. Galleries at different levels are connected by well-like pits, some of which measure two hundred and twenty-five feet from top to bottom. Through some of the lowest of these tunnels flows Echo River, still at work dissolving and wearing away the rock while on its dark way to appear at the surface as a great spring.

NATURAL BRIDGES. As a cavern enlarges and the surface of the land above it is lowered by weathering, the roof at last breaks down and the cave becomes an open ravine. A portion of the roof may for a while remain, forming a "natural bridge."

SINK HOLES. In limestone regions channels under ground may become so well developed that the water of rains rapidly drains away through them. Ground water stands low and wells must be sunk deep to find it. Little or no surface water is left to form brooks.

Thus across the limestone upland of central Kentucky one meets but three surface streams in a hundred miles. Between their valleys surface water finds its way underground by means of sink holes. These are pits, commonly funnel shaped, formed by the enlargement of crevice or joint by percolating water, or by the breakdown of some portion of the roof of a cave. By clogging of the outlet a sink hole may come to be filled by a pond.

Central Florida is a limestone region with its drainage largely subterranean and in part below the level even of the sea. Sink holes are common, and many of them are occupied by lakelets. Great springs mark the point of issue of underground streams, while some rise from beneath the sea. Silver Spring, one of the largest, discharges from a basin eight hundred feet wide and thirty feet deep a little river navigable for small steamers to its source. About the spring there are no surface streams for sixty miles.

THE KARST. Along the eastern coast of the Adriatic, as far south as Montenegro, lies a belt of limestone mountains singularly worn and honeycombed by the solvent action of water. Where forests have been cut from the mountain sides and the red soil has washed away, the surface of the white limestone forms a pathless desert of rock where each square rod has been corroded into an intricate branch work of shallow furrows and sharp ridges. Great sink holes, some of them six hundred feet deep and more, pockmark the surface of the land. The drainage is chiefly subterranean. Surface streams are rare and a portion of their courses is often under ground. Fragmentary valleys come suddenly to an end at walls of rock where the rivers which occupy the valleys plunge into dark tunnels to reappear some miles away. Ground water stands so far below the surface that it cannot be reached by wells, and the inhabitants depend on rain water stored for household uses.

The finest cavern of Europe, the Adelsberg Grotto, is in this region. Karst, the name of a part of this country, is now used to designate any region or landscape thus sculptured by the chemical action of surface and ground water. We must remember that Karst regions are rare, and striking as is the work of their subterranean streams, it is far less important than the work done by the sheets of underground water slowly seeping through all subsoils and porous rocks in other regions.

Even when gathered into definite channels, ground water does not have the erosive power of surface streams, since it carries with it little or no rock waste. Regions whose underground drainage is so perfect that the development of surface streams has been retarded or prevented escape to a large extent the leveling action of surface running waters, and may therefore stand higher than the surrounding country. The hill honeycombed by Luray Cavern, Virginia, has been attributed to this cause.

CAVERN DEPOSITS. Even in the zone of solution water may under certain circumstances deposit as well as erode. As it trickles from the roof of caverns, the lime carbonate which it has taken into solution from the layers of limestone above is deposited by evaporation in the air in icicle-like pendants called STALACTITES. As the drops splash on the floor there are built up in the same way thicker masses called STALAGMITES, which may grow to join the stalactites above, forming pillars. A stalagmitic crust often seals with rock the earth which accumulates in caverns, together with whatever relics of cave dwellers, either animals or men, it may contain.

Can you explain why slender stalactites formed by the drip of single drops are often hollow pipes?

THE ZONE OF CEMENTATION. With increasing depth subterranean water becomes more and more sluggish in its movements and more and more highly charged with minerals dissolved from the rocks above. At such depths it deposits these minerals in the pores of rocks, cementing their grains together, and in crevices and fissures, forming mineral veins. Thus below the zone of solution where the work of water is to dissolve, lies the zone of cementation where its work is chemical deposit. A part of the invisible load of waste is thus transferred from rocks near the surface to those at greater depths.

As the land surface is gradually lowered by weathering and the work of rain and streams, rocks which have lain deep within the zone of cementation are brought within the zone of solution. Thus there are exposed to view limestones, whose cracks were filled with calcite (crystallized carbonate of lime), with quartz or other minerals, and sandstones whose grains were well cemented many feet below the surface.

CAVITY FILLING. Small cavities in the rocks are often found more or less completely filled with minerals deposited from solution by water in its constant circulation underground. The process may be illustrated by the deposit of salt crystals in a cup of evaporating brine, but in the latter instance the solution is not renewed as in the case of cavities in the rocks. A cavity thus lined with inward-pointing crystals is called a GEODE.

CONCRETIONS. Ground water seeping through the pores of rocks may gather minerals disseminated throughout them into nodular masses called concretions. Thus silica disseminated through limestone is gathered into nodules of flint. While geodes grow from the outside inwards, concretions grow outwards from the center. Nor are they formed in already existing cavities as are geodes. In soft clays concretions may, as they grow, press the clay aside. In many other rocks concretions are made by the process of REPLACEMENT. Molecule by molecule the rock is removed and the mineral of the concretion substituted in its place. The concretion may in this way preserve intact the lamination lines or other structures of the rock. Clays and shales often contain concretions of lime carbonate, of iron carbonate, or of iron sulphide. Some fossil, such as a leaf or shell, frequently forms the nucleus around which the concretion grows.

Why are building stones more easily worked when "green" than after their quarry water has dried out?

DEPOSITS OF GROUND WATER IN ARID REGIONS. In arid lands where ground water is drawn by capillarity to the surface and there evaporates, it leaves as surface incrustations the minerals held in solution. White limy incrustations of this nature cover considerable tracts in northern Mexico. Evaporating beneath the surface, ground water may deposit a limy cement in beds of loose sand and gravel. Such firmly cemented layers are not uncommon in western Kansas and Nebraska, where they are known as "mortar beds."

THERMAL SPRINGS. While the lower limit of surface drainage is sea level, subterranean water circulates much below that depth, and is brought again to the surface by hydrostatic pressure. In many instances springs have a higher temperature than the average annual temperature of the region, and are then known as thermal springs. In regions of present or recent volcanic activity, such as the Yellowstone National Park, we may believe that the heat of thermal springs is derived from uncooled lavas, perhaps not far below the surface. But when hot springs occur at a distance of hundreds of miles from any volcano, as in the case of the hot springs of Bath, England, it is probable that their waters have risen from the heated rocks of the earth's interior. The springs of Bath have a temperature of 120 degrees F., 70 degrees above the average annual temperature of the place. If we assume that the rate of increase in the earth's internal heat is here the average rate, 1 degree F. to every sixty feet of descent, we may conclude that the springs of Bath rise from at least a depth of forty-two hundred feet.

Water may descend to depths from which it can never be brought back by hydrostatic pressure. It is absorbed by highly heated rocks deep below the surface. From time to time some of this deep-seated water may be returned to open air in the steam of volcanic eruptions.

SURFACE DEPOSITS OF SPRINGS. Where subterranean water returns to the surface highly charged with minerals in solution, on exposure to the air it is commonly compelled to lay down much of its invisible load in chemical deposits about the spring. These are thrown down from solution either because of cooling, evaporation, the loss of carbon dioxide, or the work of algae.

Many springs have been charged under pressure with carbon dioxide from subterranean sources and are able therefore to take up large quantities of lime carbonate from the limestone rocks through which they pass. On reaching the surface the pressure is relieved, the gas escapes, and the lime carbonate is thrown down in deposits called TRAVERTINE. The gas is sometimes withdrawn and the deposit produced in large part by the action of algae and other humble forms of plant life.

At the Mammoth Hot Springs in the valley of the Gardiner River, Yellowstone National Park, beautiful terraces and basins of travertine are now building, chiefly by means of algae which cover the bottoms, rims, and sides of the basins and deposit lime carbonate upon them in successive sheets. The rock, snow-white where dry, is coated with red and orange gelatinous mats where the algae thrive in the over-flowing waters.

Similar terraces of travertine are found to a height of fourteen hundred feet up the valley side. We may infer that the springs which formed these ancient deposits discharged near what was then the bottom of the valley, and that as the valley has been deepened by the river the ground water of the region has found lower and lower points of issue.

In many parts of the country calcareous springs occur which coat with lime carbonate mosses, twigs, and other objects over which their waters flow. Such are popularly known as petrifying springs, although they merely incrust the objects and do not convert them into stone.

Silica is soluble in alkaline waters, especially when these are hot. Hot springs rising through alkaline siliceous rocks, such as lavas, often deposit silica in a white spongy formation known as SILICEOUS SINTER, both by evaporation and by the action of algae which secrete silica from the waters. It is in this way that the cones and mounds of the geysers in the Yellowstone National Park and in Iceland have been formed.

Where water oozes from the earth one may sometimes see a rusty deposit on the ground, and perhaps an iridescent scum upon the water. The scum is often mistaken for oil, but at a touch it cracks and breaks, as oil would not do. It is a film of hydrated iron oxide, or LIMONITE, and the spring is an iron, or chalybeate, spring. Compounds of iron have been taken into solution by ground water from soil and rocks, and are now changed to the insoluble oxide on exposure to the oxygen of the air.

In wet ground iron compounds leached by ground water from the soil often collect in reddish deposits a few feet below the surface, where their downward progress is arrested by some impervious clay. At the bottom of bogs and shallow lakes iron ores sometimes accumulate to a depth of several feet.

Decaying organic matter plays a large part in these changes. In its presence the insoluble iron oxides which give color to most red and yellow rocks are decomposed, leaving the rocks of a gray or bluish color, and the soluble iron compounds which result are readily leached out,--effects seen where red or yellow clays have been bleached about some decaying tree root.

The iron thus dissolved is laid down as limonite when oxidized, as about a chalybeate spring; but out of contact with the air and in the presence of carbon dioxide supplied by decaying vegetation, as in a peat bog, it may be deposited as iron carbonate, or SIDERITE.

TOTAL AMOUNT OF UNDERGROUND WATERS. In order to realize the vast work in solution and cementation which underground waters are now doing and have done in all geological ages, we must gain some conception of their amount. At a certain depth, estimated at about six miles, the weight of the crust becomes greater than the rocks can bear, and all cavities and pores in them must be completely closed by the enormous pressure which they sustain. Below a depth of even three or four miles it is believed that ground water cannot circulate. Estimating the average pore spaces of the different rocks of the earth's crust above this depth, and the average per cents of their pore spaces occupied by water, it has been recently computed that the total amount of ground water is equal to a sheet of water one hundred feet deep, covering the entire surface of the earth.

CHAPTER III

RIVERS AND VALLEYS

THE RUN-OFF. We have traced the history of that portion of the rainfall which soaks into the ground; let us now return to that part which washes along the surface and is known as the RUN-OFF. Fed by rains and melting snows, the run-off gathers into courses, perhaps but faintly marked at first, which join more definite and deeply cut channels, as twigs their stems. In a humid climate the larger ravines through which the run-off flows soon descend below the ground-water surface. Here springs discharge along the sides of the little valleys and permanent streams begin. The water supplied by the run-off here joins that part of the rainfall which had soaked into the soil, and both now proceed together by way of the stream to the sea.

RIVER FLOODS. Streams vary greatly in volume during the year. At stages of flood they fill their immediate banks, or overrun them and inundate any low lands adjacent to the channel; at stages of low water they diminish to but a fraction of their volume when at flood.

At times of flood, rivers are fed chiefly by the run-off; at times of low water, largely or even wholly by springs.

How, then, will the water of streams differ at these times in turbidity and in the relative amount of solids carried in solution?

In parts of England streams have been known to continue flowing after eighteen months of local drought, so great is the volume of water which in humid climates is stored in the rocks above the drainage level, and so slowly is it given off in springs.

In Illinois and the states adjacent, rivers remain low in winter and a "spring freshet" follows the melting of the winter's snows. A "June rise" is produced by the heavy rains of early summer. Low water follows in July and August, and streams are again swollen to a moderate degree under the rains of autumn.

THE DISCHARGE OF STREAMS. The per cent of rainfall discharged by rivers varies with the amount of rainfall, the slope of the drainage area, the texture of the rocks, and other factors. With an annual rainfall of fifty inches in an open country, about fifty per cent is discharged; while with a rainfall of twenty inches only fifteen per cent is discharged, part of the remainder being evaporated and part passing underground beyond the drainage area. Thus the Ohio discharges thirty per cent of the rainfall of its basin, while the Missouri carries away but fifteen per cent. A number of the streams of the semi-arid lands of the West do not discharge more than five per cent of the rainfall.

Other things being equal, which will afford the larger proportion of run-off, a region underlain with granite rock or with coarse sandstone? grass land or forest? steep slopes or level land? a well-drained region or one abounding in marshes and ponds? frozen or unfrozen ground? Will there be a larger proportion of run-off after long rains or after a season of drought? after long and gentle rains, or after the same amount of precipitation in a violent rain? during the months of growing vegetation, from June to August, or during the autumn months?

DESERT STREAMS. In arid regions the ground-water surface lies so low that for the most part stream ways do not intersect it. Streams therefore are not fed by springs, but instead lose volume as their waters soak into the thirsty rocks over which they flow. They contribute to the ground water of the region instead of being increased by it. Being supplied chiefly by the run-off, they wither at times of drought to a mere trickle of water, to a chain of pools, or go wholly dry, while at long intervals rains fill their dusty beds with sudden raging torrents. Desert rivers therefore periodically shorten and lengthen their courses, withering back at times of drought for scores of miles, or even for a hundred miles from the point reached by their waters during seasons of rain.

THE GEOLOGICAL WORK OF STREAMS. The work of streams is of three kinds,--transportation, erosion, and deposition. Streams TRANSPORT the waste of the land; they wear, or ERODE, their channels both on bed and banks; and they DEPOSIT portions of their load from time to time along their courses, finally laying it down in the sea. Most of the work of streams is done at times of flood.

TRANSPORTATION

THE INVISIBLE LOAD OF STREAMS. Of the waste which a river transports we may consider first the invisible load which it carries in solution, supplied chiefly by springs but also in part by the run-off and from the solution of the rocks of its bed. More than half the dissolved solids in the water of the average river consists of the carbonates of lime and magnesia; other substances are gypsum, sodium sulphate (Glauber's salts), magnesium sulphate (Epsom salts), sodium chloride (common salt), and even silica, the least soluble of the common rock-making minerals. The amount of this invisible load is surprisingly large. The Mississippi, for example, transports each year 113,000,000 tons of dissolved rock to the Gulf.

THE VISIBLE LOAD OF STREAMS. This consists of the silt which the stream carries in suspension, and the sand and gravel and larger stones which it pushes along its bed. Especially in times of flood one may note the muddy water, its silt being kept from settling by the rolling, eddying currents; and often by placing his ear close to the bottom of a boat one may hear the clatter of pebbles as they are hurried along. In mountain torrents the rumble of bowlders as they clash together may be heard some distance away. The amount of the load which a stream can transport depends on its velocity. A current of two thirds of a mile per hour can move fine sand, while one of four miles per hour sweeps along pebbles as large as hen's eggs. The transporting power of a stream varies as the sixth power of its velocity. If its velocity is multiplied by two, its transporting power is multiplied by the sixth power of two: it can now move stones sixty-four times as large as it could before.

Stones weigh from two to three times as much as water, and in water lose the weight of the volume of water which they displace. What proportion, then, of their weight in air do stones lose when submerged?

MEASUREMENT OF STREAM LOADS. To obtain the total amount of waste transported by a river is an important but difficult matter. The amount of water discharged must first be found by multiplying the number of square feet in the average cross section of the stream by its velocity per second, giving the discharge per second in cubic feet. The amount of silt to a cubic foot of water is found by filtering samples of the water taken from different parts of the stream and at different times in the year, and drying and weighing the residues. The average amount of silt to the cubic foot of water, multiplied by the number of cubic feet of water discharged per year, gives the total load carried in suspension during that time. Adding to this the estimated amount of sand and gravel rolled along the bed, which in many swift rivers greatly exceeds the lighter material held in suspension, and adding also the total amount of dissolved solids, we reach the exceedingly important result of the total load of waste discharged by the river. Dividing the volume of this load by the area of the river basin gives another result of the greatest geological interest,-- the rate at which the region is being lowered by the combined action of weathering and erosion, or the rate of denudation.

THE RATE OF DENUDATION OF RIVER BASINS. This rate varies widely. The Mississippi basin may be taken as a representative land surface because of the varieties of surface, altitude and slope, climate, and underlying rocks which are included in its great extent. Careful measurements show that the Mississippi basin is now being lowered at a rate of one four-thousandth of a foot a year, or one foot in four thousand years. Taking this as the average rate of denudation for the land surfaces of the globe, estimates have been made of the length of time required at this rate to wash and wear the continents to the level of the sea. As the average elevation of the lands of the globe is reckoned at 2411 feet, this result would occur in nine or ten million years, if the present rate of denudation should remain unchanged. But even if no movements of the earth's crust should lift or depress the continents, the rate of wear and the removal of waste from their surfaces will not remain the same. It must constantly decrease as the lands are worn nearer to sea level and their slopes become more gentle. The length of time required to wear them away is therefore far in excess of that just stated.

The drainage area of the Potomac is 11,000 square miles. The silt brought down in suspension in a year would cover a square mile to the depth of four feet. At what rate is the Potomac basin being lowered from this cause alone?

It is estimated that the Upper Ganges is lowering its basin at the rate of one foot in 823 years, and the Po one foot in 720 years. Why so much faster than the Potomac and the Mississippi?

HOW STREAMS GET THEIR LOADS. The load of streams is derived from a number of sources, the larger part being supplied by the weathering of valley slopes. We have noticed how the mantle of waste creeps and washes to the stream ways. Watching the run-off during a rain, as it hurries muddy with waste along the gutter or washes down the hillside, we may see the beginning of the route by which the larger part of their load is delivered to rivers. Streams also secure some of their load by wearing it from their beds and banks,--a process called erosion.

EROSION

Streams erode their beds chiefly by means of their bottom load,-- the stones of various sizes and the sand and even the fine mud which they sweep along. With these tools they smooth, grind, and rasp the rock of their beds, using them in much the fashion of sandpaper or a file.

WEATHERING OF RIVER BEDS. The erosion of stream beds is greatly helped by the work of the weather. Especially at low water more or less of the bed is exposed to the action of frost and heat and cold, joints are opened, rocks are pried loose and broken up and made ready to be swept away by the stream at time of flood.

POTHOLES. In rapids streams also drill out their rocky beds. Where some slight depression gives rise to an eddy, the pebbles which gather in it are whirled round and round, and, acting like the bit of an auger, bore out a cylindrical pit called a pothole. Potholes sometimes reach a depth of a score of feet. Where they are numerous they aid materially in deepening the channel, as the walls between them are worn away and they coalesce.

WATERFALLS. One of the most effective means of erosion which the river possesses is the waterfall. The plunging water dislodges stones from the face of the ledge over which it pours, and often undermines it by excavating a deep pit at its base. Slice after slice is thus thrown down from the front of the cliff, and the cataract cuts its way upstream leaving a gorge behind it.

NIAGARA FALLS. The Niagara River flows from Lake Erie at Buffalo in a broad channel which it has cut but a few feet below the level of the region. Some thirteen miles from the outlet it plunges over a ledge one hundred and seventy feet high into the head of a narrow gorge which extends for seven miles to the escarpment of the upland in which the gorge is cut. The strata which compose the upland dip gently upstream and consist at top of a massive limestone, at the Falls about eighty feet thick, and below of soft and easily weathered shale. Beneath the Falls the underlying shale is cut and washed away by the descending water and retreats also because of weathering, while the overhanging limestone breaks down in huge blocks from time to time.

Niagara is divided by Goat Island into the Horseshoe Falls and the American Falls. The former is supplied by the main current of the river, and from the semicircular sweep of its rim a sheet of water in places at least fifteen or twenty feet deep plunges into a pool a little less than two hundred feet in depth. Here the force of the falling water is sufficient to move about the fallen blocks of limestone and use them in the excavation of the shale of the bed. At the American Falls the lesser branch of the river, which flows along the American side of Goat Island, pours over the side of the gorge and breaks upon a high talus of limestone blocks which its smaller volume of water is unable to grind to pieces and remove.

A series of surveys have determined that from 1842 to 1890 the Horseshoe Falls retreated at the rate of 2.18 feet per year, while the American Falls retreated at the rate of 0.64 feet in the same period. We cannot doubt that the same agency which is now lengthening the gorge at this rapid rate has cut it back its entire length of seven miles.

While Niagara Falls have been cutting back a gorge seven miles long and from two hundred to three hundred feet deep, the river above the Falls has eroded its bed scarcely below the level of the upland on which it flows. Like all streams which are the outlets of lakes, the Niagara flows out of Lake Erie clear of sediment, as from a settling basin, and carries no tools with which to abrade its bed. We may infer from this instance how slight is the erosive power of clear water on hard rock.

Assuming that the rate of recession of the combined volumes of the American and Horseshoe Falls was three feet a year below Goat Island, and ASSUMING THAT THIS RATE HAS BEEN UNIFORM IN THE PAST, how long is it since the Niagara River fell over the edge of the escarpment where now is the mouth of the present gorge?

The profile of the bed of the Niagara along the gorge (Fig. 39) shows alternating deeps and shallows which cannot be accounted for, except in a single instance, by the relative hardness of the rocks of the river bed. The deeps do not exceed that at the foot of the Horseshoe Falls at the present time. When the gorge was being cut along the shallows, how did the Falls compare in excavating power, in force, and volume with the Niagara of to-day? How did the rate of recession at those times compare with the present rate? Is the assumption made above that the rate of recession has been uniform correct?

The first stretch of shallows below the Falls causes a tumultuous rapid impossible to sound. Its depth has been estimated at thirty- five feet. From what data could such an estimate be made?

Suggest a reason why the Horseshoe Falls are convex upstream.

At the present rate of recession which will reach the head of Goat Island the sooner, the American or the Horseshoe Falls? What will be the fate of the Falls left behind when the other has passed beyond the head of the island?

The rate at which a stream erodes its bed depends in part upon the nature of the rocks over which it flows. Will a stream deepen its channel more rapidly on massive or on thin-bedded and close-jointed rocks? on horizontal strata or on strata steeply inclined?

DEPOSITION

While the river carries its invisible load of dissolved rock on without stop to the sea, its load of visible waste is subject to many delays en route. Now and again it is laid aside, to be picked up later and carried some distance farther on its way. One of the most striking features of the river therefore is the waste accumulated along its course, in bars and islands in the channel, beneath its bed, and in flood plains along its banks. All this alluvium, to use a general term for river deposits, with which the valley is cumbered is really en route to the sea; it is only temporarily laid aside to resume its journey later on. Constantly the river is destroying and rebuilding its alluvial deposits, here cutting and there depositing along its banks, here eroding and there building a bar, here excavating its bed and there filling it up, and at all times carrying the material picked up at one point some distance on downstream before depositing it at another.

These deposits are laid down by slackening currents where the velocity of the stream is checked, as on the inner side of curves, and where the slope of the bed is diminished, and in the lee of islands, bridge piers and projecting points of land. How slight is the check required to cause a current to drop a large part of its load may be inferred from the law of the relation of the transporting power to the velocity. If the velocity is decreased one half, the current can move fragments but one sixty-fourth the size of those which it could move before, and must drop all those of larger size.

Will a river deposit more at low water or at flood? when rising or when falling?

STRATIFICATION. River deposits are stratified, as may be seen in any fresh cut in banks or bars. The waste of which they are built has been sorted and deposited in layers, one above another; some of finer and some of coarser material. The sorting action of running water depends on the fact that its transporting power varies with the velocity. A current whose diminishing velocity compels it to drop coarse gravel, for example, is still able to move all the finer waste of its load, and separating it from the gravel, carries it on downstream; while at a later time slower currents may deposit on the gravel bed layers of sand, and, still later, slack water may leave on these a layer of mud. In case of materials lighter than water the transporting power does not depend on the velocity, and logs of wood, for instance, are floated on to the sea on the slowest as well as on the most rapid currents.

CROSS BEDDING. A section of a bar exposed at low water may show that it is formed of layers of sand, or coarser stuff, inclined downstream as steeply often as the angle of repose of the material. From a boat anchored over the lower end of a submerged sand bar we may observe the way in which this structure, called cross bedding, is produced. Sand is continually pushed over the edge of the bar at b (Fig. 42) and comes to rest in successive layers on the sloping surface. At the same time the bar may be worn away at the upper end, a, and thus slowly advance down stream. While the deposit is thus cross bedded, it constitutes as a whole a stratum whose upper and lower surfaces are about horizontal. In sections of river banks one may often see a vertical succession of cross-bedded strata, each built in the way described.

WATER WEAR. The coarser material of river deposits, such as cobblestones, gravel, and the larger grains of sand, are WATER WORN, or rounded, except when near their source. Rolling along the bottom they have been worn round by impact and friction as they rubbed against one another and the rocky bed of the stream.

Experiments have shown that angular fragments of granite lose nearly half their weight and become well rounded after traveling fifteen miles in rotating cylinders partly filled with water. Marbles are cheaply made in Germany out of small limestone cubes set revolving in a current of water between a rotating bed of stone and a block of oak, the process requiring but about fifteen minutes. It has been found that in the upper reaches of mountain streams a descent of less than a mile is sufficient to round pebbles of granite.

LAND FORMS DUE TO RIVER EROSION

RIVER VALLEYS. In their courses to the sea, rivers follow valleys of various forms, some shallow and some deep, some narrow and some wide. Since rivers are known to erode their beds and banks, it is a fair presumption that, aided by the weather, they have excavated the valleys in which they flow.

Moreover, a bird's-eye view or a map of a region shows the significant fact that the valleys of a system unite with one another in a branch work, as twigs meet their stems and the branches of a tree its trunk. Each valley, from that of the smallest rivulet to that of the master stream, is proportionate to the size of the stream which occupies it. With a few explainable exceptions the valleys of tributaries join that of the trunk stream at a level; there is no sudden descent or break in the bed at the point of juncture. These are the natural consequences which must follow if the land has long been worked upon by streams, and no other process has ever been suggested which is competent to produce them. We must conclude that valley systems have been formed by the river systems which drain them, aided by the work of the weather; they are not gaping fissures in the earth's crust, as early observers imagined, but are the furrows which running water has drawn upon the land.

As valleys are made by the slow wear of streams and the action of the weather, they pass in their development through successive stages, each of which has its own characteristic features. We may therefore classify rivers and valleys according to the stage which they have reached in their life history from infancy to old age.

YOUNG RIVER VALLEYS

INFANCY. The Red River of the North. A region in northwestern Minnesota and the adjacent portions of North Dakota and Manitoba was so recently covered by the waters of an extinct lake, known as Lake Agassiz, that the surface remains much as it was left when the lake was drained away. The flat floor, spread smooth with lake-laid silts, is still a plain, to the eye as level as the sea. Across it the Red River of the North and its branches run in narrow, ditch-like channels, steep-sided and shallow, not exceeding sixty feet in depth, their gradients differing little from the general slopes of the region. The trunk streams have but few tributaries; the river system, like a sapling with few limbs, is still undeveloped. Along the banks of the trunk streams short gullies are slowly lengthening headwards, like growing twigs which are sometime to become large branches.

The flat interstream areas are as yet but little scored by drainage lines, and in wet weather water lingers in ponds in any initial depressions on the plain.

CONTOURS. In order to read the topographic maps of the text-book and the laboratory the student should know that contours are lines drawn on maps to represent relief, all points on any given contour being of equal height above sea level. The CONTOUR INTERVAL is the uniform vertical distance between two adjacent contours and varies on different maps.

To express regions of faint relief a contour interval of ten or twenty feet is commonly selected; while in mountainous regions a contour interval of two hundred and fifty, five hundred, or even one thousand feet may be necessary in order that the contours may not be too crowded for easy reading.

Whether a river begins its life on a lake plain, as in the example just cited, or upon a coastal plain lifted from beneath the sea or on a spread of glacial drift left by the retreat of continental ice sheets, such as covers much of Canada and the northeastern parts of the United States, its infantile stage presents the same characteristic features,--a narrow and shallow valley, with undeveloped tributaries and undrained interstream areas. Ground water stands high, and, exuding in the undrained initial depressions, forms marshes and lakes.

LAKES. Lakes are perhaps the most obvious of these fleeting features of infancy. They are short-lived, for their destruction is soon accomplished by several means. As a river system advances toward maturity the deepening and extending valleys of the tributaries lower the ground-water surface and invade the undrained depressions of the region. Lakes having outlets are drained away as their basin rims are cut down by the outflowing streams,--a slow process where the rim is of hard rock, but a rapid one where it is of soft material such as glacial drift.

Lakes are effaced also by the filling of their basins. Inflowing streams and the wash of rains bring in waste. Waves abrade the shore and strew the debris worn from it over the lake bed. Shallow lakes are often filled with organic matter from decaying vegetation.

Does the outflowing stream, from a lake carry sediment? How does this fact affect its erosive power on hard rock? on loose material?

Lake Geneva is a well-known example of a lake in process of obliteration. The inflowing Rhone has already displaced the waters of the lake for a length of twenty miles with the waste brought down from the high Alps. For this distance there extends up the Rhone Valley an alluvial plain, which has grown lakeward at the rate of a mile and a half since Roman times, as proved by the distance inland at which a Roman port now stands.

How rapidly a lake may be silted up under exceptionally favorable conditions is illustrated by the fact that over the bottom of the artificial lake, of thirty-five square miles, formed behind the great dam across the Colorado River at Austin, Texas, sediments thirty-nine feet deep gathered in seven years.

Lake Mendota, one of the many beautiful lakes of southern Wisconsin, is rapidly cutting back the soft glacial drift of its shores by means of the abrasion of its waves. While the shallow basin is thus broadened, it is also being filled with the waste; and the time is brought nearer when it will be so shoaled that vegetation can complete the work of its effacement.

Along the margin of a shallow lake mosses, water lilies, grasses, and other water-loving plants grow luxuriantly. As their decaying remains accumulate on the bottom, the ring of marsh broadens inwards, the lake narrows gradually to a small pond set in the midst of a wide bog, and finally disappears. All stages in this process of extinction may be seen among the countless lakelets which occupy sags in the recent sheets of glacial drift in the northern states; and more numerous than the lakes which still remain are those already thus filled with carbonaceous matter derived from the carbon dioxide of the atmosphere. Such fossil lakes are marked by swamps or level meadows underlain with muck.

THE ADVANCE TO MATURITY. The infantile stage is brief. As a river advances toward maturity the initial depressions, the lake basins of its area, are gradually effaced. By the furrowing action of the rain wash and the head ward lengthening, of tributaries a branchwork of drainage channels grows until it covers the entire area, and not an acre is left on which the fallen raindrop does not find already cut for it an uninterrupted downward path which leads it on by way of gully, brook, and river to the sea. The initial surface of the land, by whatever agency it was modeled, is now wholly destroyed; the region is all reduced to valley slopes.

THE LONGITUDINAL PROFILE OF A STREAM. This at first corresponds with the initial surface of the region on which the stream begins to flow, although its way may lead through basins and down steep descents. The successive profiles to which it reduces its bed are illustrated in Figure 51. As the gradient, or rate of descent of its bed, is lowered, the velocity of the river is decreased until its lessening energy is wholly consumed in carrying its load and it can no longer erode its bed. The river is now AT GRADE, and its capacity is just equal to its load. If now its load is increased the stream deposits, and thus builds up, or AGGRADES, its bed. On the other hand, if its load is diminished it has energy to spare, and resuming its work of erosion, DEGRADES its bed. In either case the stream continues aggrading or degrading until a new gradient is found where the velocity is just sufficient to move the load, and here again it reaches grade.

V-VALLEYS. Vigorous rivers well armed with waste make short work of cutting their beds to grade, and thus erode narrow, steep-sided gorges only wide enough at the base to accommodate the stream. The steepness of the valley slopes depends on the relative rates at which the bed is cut down by the stream and the sides are worn back by the weather. In resistant rock a swift, well-laden stream may saw out a gorge whose sides are nearly or even quite vertical, but as a rule young valleys whose streams have not yet reached grade are V-shaped; their sides flare at the top because here the rocks have longest been opened up to the action of the weather. Some of the deepest canyons may be found where a rising land mass, either mountain range or plateau, has long maintained by its continued uplift the rivers of the region above grade.

In the northern hemisphere the north sides of river valleys are sometimes of more gentle slope than the south sides. Can you suggest a reason?

THE GRAND CANYON OF THE COLORADO RIVER IN ARIZONA. The Colorado River trenches the high plateau of northern Arizona with a colossal canyon two hundred and eighteen miles long and more than a mile in greatest depth. The rocks in which the canyon is cut are for the most part flat-lying, massive beds of limestones and sandstones, with some shales, beneath which in places harder crystalline rocks are disclosed. Where the canyon is deepest its walls have been profoundly dissected. Lateral ravines have widened into immense amphitheaters, leaving between them long ridges of mountain height, buttressed and rebuttressed with flanking spurs and carved into majestic architectural forms. From the extremity of one of these promontories it is two miles or more across the gulf to the point of the one opposite, and the heads of the amphitheaters are thirteen miles apart.

The lower portion of the canyon is much narrower (Fig. 54) and its walls of dark crystalline rock sink steeply to the edge of the river, a swift, powerful stream a few hundred feet wide, turbid with reddish silt, by means of which it continually rasps its rocky bed as it hurries on. The Colorado is still deepening its gorge. In the Grand Canyon its gradient is seven and one half feet to the mile, but, as in all ungraded rivers, the descent is far from uniform. Graded reaches in soft rock alternate with steeper declivities in hard rock, forming rapids such as, for example, a stretch of ten miles where the fall averages twenty-one feet to the mile. Because of these dangerous rapids the few exploring parties who have traversed the Colorado canyon have done so at the hazard of their lives.

The canyon has been shaped by several agencies. Its depth is due to the river which has sawed its way far toward the base of a lofty rising plateau. Acting alone this would have produced a slitlike gorge little wider than the breadth of the stream. The impressive width of the canyon and the magnificent architectural masses which fill it are owing to two causes.: Running water has gulched the walls and weathering has everywhere attacked and driven them back. The horizontal harder beds stand out in long lines of vertical cliffs, often hundreds of feet in height, at whose feet talus slopes conceal the outcrop of the weaker strata. As the upper cliffs have been sapped and driven back by the weather, broad platforms are left at their bases and the sides of the canyon descend to the river by gigantic steps. Far up and down the canyon the eye traces these horizontal layers, like the flutings of an elaborate molding, distinguishing each by its contour as well as by its color and thickness.

The Grand Canyon of the Colorado is often and rightly cited as an example of the stupendous erosion which may be accomplished by a river. And yet the Colorado is a young stream and its work is no more than well begun. It has not yet wholly reached grade, and the great task of the river and its tributaries--the task of leveling the lofty plateau to a low plain and of transporting it grain by grain to the sea--still lies almost entirely in the future.

WATERFALLS AND RAPIDS. Before the bed of a stream is reduced to grade it may be broken by abrupt descents which give rise to waterfalls and rapids. Such breaks in a river's bed may belong to the initial surface over which it began its course; still more commonly are they developed in the rock mass through which it is cutting its valley. Thus, wherever a stream leaves harder rocks to flow over softer ones the latter are quickly worn below the level of the former, and a sharp change in slope, with a waterfall or rapid, results.

At time of flood young tributaries with steeper courses than that of the trunk stream may bring down stones and finer waste, which the gentler current cannot move along, and throw them as a dam across its way. The rapids thus formed are also ephemeral, for as the gradient of the tributaries is lowered the main stream becomes able to handle the smaller and finer load which they discharge.

A rare class of falls is produced where the minor tributaries of a young river are not able to keep pace with their master stream in the erosion of their beds because of their smaller volume, and thus join it by plunging over the side of its gorge. But as the river approaches grade and slackens its down cutting, the tributaries sooner or later overtake it, and effacing their falls, unite with it on a level.

Waterfalls and rapids of all kinds are evanescent features of a river's youth. Like lakes they are soon destroyed, and if any long time had already elapsed since their formation they would have been obliterated already.

LOCAL BASELEVELS. That balanced condition called grade, where a river neither degrades its bed by erosion nor aggrades it by deposition, is first attained along reaches of soft rocks, ungraded outcrops of hard rocks remaining as barriers which give rise to rapids or falls. Until these barriers are worn away they constitute local baselevels, below which level the stream, up valley from them, cannot cut. They are eroded to grade one after another, beginning with the least strong, or the one nearest the mouth of the stream. In a similar way the surface of a lake in a river's course constitutes for all inflowing streams a local baselevel, which disappears when the basin is filled or drained.

MATURE AND OLD RIVERS

Maturity is the stage of a river's complete development and most effective work. The river system now has well under way its great task of wearing down the land mass which it drains and carrying it particle by particle to the sea. The relief of the land is now at its greatest; for the main channels have been sunk to grade, while the divides remain but little worn below their initial altitudes. Ground water now stands low. The run-off washes directly to the streams, with the least delay and loss by evaporation in ponds and marches; the discharge of the river is therefore at its height. The entire region is dissected by stream ways. The area of valley slopes is now largest and sheds to the streams a heavier load of waste than ever before. At maturity the river system is doing its greatest amount of work both in erosion and in the carriage of water and of waste to the sea.

LATERAL EROSION. On reaching grade a river ceases to scour its bed, and it does not again begin to do so until some change in load or volume enables it to find grade at a lower level. On the other hand, a stream erodes its banks at all stages in its history, and with graded rivers this process, called lateral erosion, or PLANATION, is specially important. The current of a stream follows the outer side of all curves or bends in the channel, and on this side it excavates its bed the deepest and continually wears and saps its banks. On the inner side deposition takes place in the more shallow and slower-moving water. The inner bank of bends is thus built out while the outer bank is worn away. By swinging its curves against the valley sides a graded river continually cuts a wider and wider floor. The V-valley of youth is thus changed by planation to a flat-floored valley with flaring sides which gradually become subdued by the weather to gentle slopes. While widening their valleys streams maintain a constant width of channel, so that a wide-floored valley does not signify that it ever was occupied by a river of equal width.

THE GRADIENT. The gradients of graded rivers differ widely. A large river with a light load reaches grade on a faint slope, while a smaller stream heavily burdened with waste requires a steep slope to give it velocity sufficient to move the load.

The Platte, a graded river of Nebraska with its headwaters in the Rocky Mountains, is enfeebled by the semi-arid climate of the Great Plains and surcharged with the waste brought down both by its branches in the mountains and by those whose tracks lie over the soft rocks of the plains. It is compelled to maintain a gradient of eight feet to the mile in western Nebraska. The Ohio reaches grade with a slope of less than four inches to the mile from Cincinnati to its mouth, and the powerful Mississippi washes along its load with a fall of but three inches per mile from Cairo to the Gulf.

Other things being equal, which of graded streams will have the steeper gradient, a trunk stream or its tributaries? a stream supplied with gravel or one with silt?

Other factors remaining the same, what changes would occur if the Platte should increase in volume? What changes would occur if the load should be increased in amount or in coarseness?

THE OLD AGE OF RIVERS. As rivers pass their prime, as denudation lowers the relief of the region, less waste and finer is washed over the gentler slopes of the lowering hills. With smaller loads to carry, the rivers now deepen their valleys and find grade with fainter declivities nearer the level of the sea. This limit of the level of the sea beneath which they cannot erode is known as baselevel. [Footnote: The term "baselevel" is also used to designate the close approximation to sea level to which streams are able to subdue the land.] As streams grow old they approach more and more closely to baselevel, although they are never able to attain it. Some slight slope is needed that water may flow and waste be transported over the land. Meanwhile the relief of the land has ever lessened. The master streams and their main tributaries now wander with sluggish currents over the broad valley floors which they have planed away; while under the erosion of their innumerable branches and the wear of the weather the divides everywhere are lowered and subdued to more and more gentle slopes. Mountains and high plateaus are thus reduced to rolling hills, and at last to plains, surmounted only by such hills as may still be unreduced to the common level, because of the harder rocks of which they are composed or because of their distance from the main erosion channels. Such regions of faint relief, worn down to near base level by subaerial agencies, are known as PENEPLAINS (almost plains). Any residual masses which rise above them are called MONADNOCKS, from the name of a conical peak of New Hampshire which overlooks the now uplifted peneplain of southern New England.

In its old age a region becomes mantled with thick sheets of fine and weathered waste, slowly moving over the faint slopes toward the water ways and unbroken by ledges of bare rock. In other words, the waste mantle also is now graded, and as waterfalls have been effaced in the river beds, so now any ledges in the wide streams of waste are worn away and covered beneath smooth slopes of fine soil. Ground water stands high and may exude in areas of swamp. In youth the land mass was roughhewn and cut deep by stream erosion. In old age the faint reliefs of the land dissolve away, chiefly under the action of the weather, beneath their cloak of waste.

THE CYCLE OF EROSION. The successive stages through which a land mass passes while it is being leveled to the sea constitute together a cycle of erosion. Each stage of the cycle from infancy to old age leaves, as we have seen, its characteristic records in the forms sculptured on the land, such as the shapes of valleys and the contours of hills and plains. The geologist is thus able to determine by the land forms of any region the stage in the erosion cycle to which it now belongs, and knowing what are the earlier stages of the cycle, to read something of the geological history of the region.

INTERRUPTED CYCLES. So long a time is needed to reduce a land mass to baselevel that the process is seldom if ever completed during a single uninterrupted cycle of erosion. Of all the various interruptions which may occur the most important are gradual movements of the earth's crust, by which a region is either depressed or elevated relative to sea level.

The DEPRESSION of a region hastens its old age by decreasing the gradient of streams, by destroying their power to excavate their beds and carry their loads to a degree corresponding to the amount of the depression, and by lessening the amount of work they have to do. The slackened river currents deposit their waste in flood plains which increase in height as the subsidence continues. The lower courses of the rivers are invaded by the sea and become estuaries, while the lower tributaries are cut off from the trunk stream.

ELEVATION, on the other hand, increases the activity of all agencies of weathering, erosion, and transportation, restores the region to its youth, and inaugurates a new cycle of erosion. Streams are given a steeper gradient, greater velocity, and increased energy to carry their loads and wear their beds.

They cut through the alluvium of their flood plains, leaving it on either bank as successive terraces, and intrench themselves in the underlying rock. In their older and wider valleys they cut narrow, steep-walled inner gorges, in which they flow swiftly over rocky floors, broken here and there by falls and rapids where a harder layer of rock has been discovered. Winding streams on plains may thus incise their meanders in solid rock as the plains are gradually uplifted. Streams which are thus restored to their youth are said to be REVIVED.

As streams cut deeper and the valley slopes are steepened, the mantle of waste of the region undergoing elevation is set in more rapid movement. It is now removed particle by particle faster than it forms. As the waste mantle thins, weathering attacks the rocks of the region more energetically until an equilibrium is reached again; the rocks waste rapidly and their waste is as rapidly removed.

DISSECTED PENEPLAINS. When a rise of the land brings one cycle to an end and begins another, the characteristic land forms of each cycle are found together and the topography of the region is composite until the second cycle is so far advanced that the land forms of the first cycle are entirely destroyed. The contrast between the land surfaces of the later and the earlier cycles is most striking when the earlier had advanced to age and the later is still in youth. Thus many peneplains which have been elevated and dissected have been recognized by the remnants of their ancient erosion surfaces, and the length of time which has elapsed since their uplift has been measured by the stage to which the new cycle has advanced.

THE PIEDMONT BELT. As an example of an ancient peneplain uplifted and dissected we may cite the Piedmont Belt, a broad upland lying between the Appalachian Mountains and the Atlantic coastal plain. The surface of the Piedmont is gently rolling. The divides, which are often smooth areas of considerable width, rise to a common plane, and from them one sees in every direction an even sky line except where in places some lone hill or ridge may lift itself above the general level (Fig. 62). The surface is an ancient one, for the mantle of residual waste lies deep upon it, soils are reddened by long oxidation, and the rocks are rotted to a depth of scores of feet.

At present, however, the waste mantle is not forming so rapidly as it is being removed. The streams of the upland are actively engaged in its destruction. They flow swiftly in narrow, rock- walled valleys over rocky beds. This contrast between the young streams and the aged surface which they are now so vigorously dissecting can only be explained by the theory that the region once stood lower than at present and has recently been upraised. If now we imagine the valleys refilled with the waste which the streams have swept away, and the upland lowered, we restore the Piedmont region to the condition in which it stood before its uplift and dissection,--a gently rolling plain, surmounted here and there by isolated hills and ridges.

The surface of the ancient Piedmont plain, as it may be restored from the remnants of it found on the divides, is not in accordance with the structures of the country rocks. Where these are exposed to view they are seen to be far from horizontal. On the walls of river gorges they dip steeply and in various directions and the streams flow over their upturned edges. As shown in Figure 67, the rocks of the Piedmont have been folded and broken and tilted.

It is not reasonable to believe that when the rocks of the Piedmont were thus folded and otherwise deformed the surface of the region was a plain. The upturned layers have not always stopped abruptly at the even surface of the Piedmont plain which now cuts across them. They are the bases of great folds and tilted blocks which must once have risen high in air. The complex and disorderly structures of the Piedmont rocks are those seen in great mountain ranges, and there is every reason to believe that these rocks after their deformation rose to mountain height.

The ancient Piedmont plain cuts across these upturned rocks as independently of their structure as the even surface of the sawed stump of some great tree is independent of the direction of its fibers. Hence the Piedmont plain as it was before its uplift was not a coastal plain formed of strata spread in horizontal sheets beneath the sea and then uplifted; nor was it a structural plain, due to the resistance to erosion of some hard, flat-lying layer of rock. Even surfaces developed on rocks of discordant structure, such as the Piedmont shows, are produced by long denudation, and we may consider the Piedmont as a peneplain formed by the wearing down of mountain ranges, and recently uplifted.

THE LAURENTIAN PENEPLAIN. This is the name given to a denuded surface on very ancient rocks which extends from the Arctic Ocean to the St. Lawrence River and Lake Superior, with small areas also in northern Wisconsin and New York. Throughout this U-shaped area, which incloses Hudson Bay within its arms, the country rocks have the complicated and contorted structures which characterize mountain ranges. But the surface of the area is by no means mountainous. The sky line when viewed from the divides is unbroken by mountain peaks or rugged hills. The surface of the arm west of Hudson Bay is gently undulating and that of the eastern arm has been roughened to low-rolling hills and dissected in places by such deep river gorges as those of the Ottawa and Saguenay. This immense area may be regarded as an ancient peneplain truncating the bases of long-vanished mountains and dissected after elevation.

In the examples cited the uplift has been a broad one and to comparatively little height. Where peneplains have been uplifted to great height and have since been well dissected, and where they have been upfolded and broken and uptilted, their recognition becomes more difficult. Yet recent observers have found evidences of ancient lowland surfaces of erosion on the summits of the Allegheny ridges, the Cascade Mountains (Fig. 69), and the western slope of the Sierra Nevadas.

THE SOUTHERN APPALACHIAN REGION. We have here an example of an area the latter part of whose geological history may be deciphered by means of its land forms. The generalized section of Figure 70, which passes from west to east across a portion of the region in eastern Tennessee, shows on the west a part of the broad Cumberland plateau. On the east is a roughened upland platform, from which rise in the distance the peaks of the Great Smoky Mountains. The plateau, consisting of strata but little changed from their original flat-lying attitude, and the platform, developed on rocks of disordered structure made crystalline by heat and pressure, both stand at the common level of the line AB. They are separated by the Appalachian valley, forty miles wide, cut in strata which have been folded and broken into long narrow blocks. The valley is traversed lengthwise by long, low ridges, the outcropping edges of the harder strata, which rise to about the same level,--that of the line cd. Between these ridges stretch valley lowlands at the level ef excavated in the weaker rocks, while somewhat below them lie the channels of the present streams now busily engaged in deepening their beds.

THE VALLEY LOWLANDS. Were they planed by graded or ungraded streams? Have the present streams reached grade? Why did the streams cease widening the floors of the valley lowlands? How long since? When will they begin anew the work of lateral planation? What effect will this have on the ridges if the present cycle of erosion continues long uninterrupted?

THE RIDGES OF THE APPALACHIAN VALLEY. Why do they stand above the valley lowlands? Why do their summits lie in about the same plane? Refilling the valleys intervening between these ridges with the material removed by the streams, what is the nature of the surface thus restored? Does this surface cd accord with the rock structures on which' it has been developed? How may it have been made? At what height did the land stand then, compared with its present height? What elevations stood above the surface cd? Why? What name may you use to designate them? How does the length of time needed to develop the surface cd compare with that needed to develop the valley lowlands?

THE PLATFORM AND PLATEAU. Why do they stand at a common level ab? Of what surface may they be remnants? Is it accordant with the rock structure? How was it produced? What unconsumed masses overlooked it? Did the rocks of the Appalachian valley stand above this surface when it was produced? Did they then stand below it? Compare the time needed to develop this surface with that needed to develop cd. Which surface is the older?

How many cycles of erosion are represented here? Give the erosion history of the region by cycles, beginning with the oldest, the work done in each and the work left undone, what brought each cycle to a close, and how long relatively it continued.

CHAPTER IV

RIVER DEPOSITS

The characteristic features of river deposits and the forms which they assume may be treated under three heads: (1) valley deposits, (2) basin deposits, and (3) deltas.

VALLEY DEPOSITS

FLOOD PLAINS are the surfaces of the alluvial deposits which streams build along their courses at times of flood. A swift current then sweeps along the channel, while a shallow sheet of water moves slowly over the flood plain, spreading upon it a thin layer of sediment. It has been estimated that each inundation of the Nile leaves a layer of fertilizing silt three hundredths of an inch thick over the flood plain of Egypt.

Flood plains may consist of a thin spread of alluvium over the flat rock floor of a valley which is being widened by the lateral erosion of a graded stream (Fig. 60). Flood-plain deposits of great thickness may be built by aggrading rivers even in valleys whose rock floors have never been thus widened.

A cross section of a flood plain shows that it is highest next the river, sloping gradually thence to the valley sides. These wide natural embankments are due to the fact that the river deposit is heavier near the bank, where the velocity of the silt-laden channel current is first checked by contact with the slower-moving overflow.

Thus banked off from the stream, the outer portions of a flood plain are often ill-drained and swampy, and here vegetal deposits, such as peat, may be interbedded with river silts.

A map of a wide flood plain, such as that of the Mississippi or the Missouri (Fig. 77), shows that the courses of the tributaries on entering it are deflected downstream. Why?

The aggrading streams by which flood plains are constructed gradually build their immediate banks and beds to higher and higher levels, and therefore find it easy at times of great floods to break their natural embankments and take new courses over the plain. In this way they aggrade each portion of it in turn by means of their shifting channels,

BRAIDED CHANNELS. A river actively engaged in aggrading its valley with coarse waste builds a flood plain of comparatively steep gradient and often flows down it in a fairly direct course and through a network of braided channels. From time to time a channel becomes choked with waste, and the water no longer finding room in it breaks out and cuts and builds itself a new way which reunites down valley with the other channels. Thus there becomes established a network of ever-changing channels inclosing low islands of sand and gravel.

TERRACES. While aggrading streams thus tend to shift their channels, degrading streams, on the contrary, become more and more deeply intrenched in their valleys. It often occurs that a stream, after having built a flood plain, ceases to aggrade its bed because of a lessened load or for other reasons, such as an uplift of the region, and begins instead to degrade it. It leaves the original flood plain out of reach of even the highest floods. When again it reaches grade at a lower level it produces a new flood plain by lateral erosion in the older deposits, remnants of which stand as terraces on one or both sides of the valley. In this way a valley may be lined with a succession of terraces at different levels, each level representing an abandoned flood plain.

MEANDERS. Valleys aggraded with fine waste form well-nigh level plains over which streams wind from side to side of a direct course in symmetric bends known as meanders, from the name of a winding river of Asia Minor. The giant Mississippi has developed meanders with a radius of one and one half miles, but a little creek may display on its meadow as perfect curves only a rod or so in radius. On the flood plain of either river or creek we may find examples of the successive stages in the development of the meander, from its beginning in the slight initial bend sufficient to deflect the current against the outer side. Eroding here and depositing on the inner side of the bend, it gradually reaches first the open bend whose width and length are not far from equal, and later that of the horseshoe meander whose diameter transverse to the course of the stream is much greater than that parallel with it. Little by little the neck of land projecting into the bend is narrowed, until at last it is cut through and a "cut-off" is established. The old channel is now silted up at both ends and becomes a crescentic lagoon, or oxbow lake, which fills gradually to an arc-shaped shallow depression.

FLOOD PLAINS CHARACTERISTIC OF MATURE RIVERS. On reaching grade a stream planes a flat floor for its continually widening valley. Ever cutting on the outer bank of its curves, it deposits on the inner bank scroll-like flood-plain patches. For a while the valley bluffs do not give its growing meanders room to develop to their normal size, but as planation goes on, the bluffs are driven back to the full width of the meander belt and still later to a width which gives room for broad stretches of flood plain on either side.

Usually a river first attains grade near its mouth, and here first sinks its bed to near baselevel. Extending its graded course upstream by cutting away barrier after barrier, it comes to have a widened and mature valley over its lower course, while its young headwaters are still busily eroding their beds. Its ungraded branches may thus bring down to its lower course more waste than it is competent to carry on to the sea, and here it aggrades its bed and builds a flood plain in order to gain a steeper gradient and velocity enough to transport its load.

As maturity is past and the relief of the land is lessened, a smaller and smaller load of waste is delivered to the river. It now has energy to spare and again degrades its valley, excavating its former flood plains and leaving them in terraces on either side, and at last in its old age sweeping them away.

ALLUVIAL CONES AND FANS. In hilly and mountainous countries one often sees on a valley side a conical or fan-shaped deposit of waste at the mouth of a lateral stream. The cause is obvious: the young branch has not been able as yet to wear its bed to accordant level with the already deepened valley of the master stream. It therefore builds its bed to grade at the point of juncture by depositing here its load of waste,--a load too heavy to be carried along the more gentle profile of the trunk valley.

Where rivers descend from a mountainous region upon the plain they may build alluvial fans of exceedingly gentle slope. Thus the rivers of the western side of the Sierra Nevada Mountains have spread fans with a radius of as much as forty miles and a slope too slight to be detected without instruments, where they leave the rock-cut canyons in the mountains and descend upon the broad central valley of California.

As a river flows over its fan it commonly divides into a branchwork of shifting channels called DISTRIBUTARIES, since they lead off the water from the main stream. In this way each part of the fan is aggraded and its symmetric form is preserved.

PIEDMONT PLAINS. Mountain streams may build their confluent fans into widespread piedmont (foot of the mountain) alluvial plains. These are especially characteristic of arid lands, where the streams wither as they flow out upon the thirsty lowlands and are therefore compelled to lay down a large portion of their load. In humid climates mountain-born streams are usually competent to carry their loads of waste on to the sea, and have energy to spare to cut the lower mountain slopes into foothills. In arid regions foothills are commonly absent and the ranges rise, as from pedestals, above broad, sloping plains of stream-laid waste.

THE HIGH PLAINS. The rivers which flow eastward from the Rocky Mountains have united their fans in a continuous sheet of waste which stretches forward from the base of the mountains for hundreds of miles and in places is five hundred feet thick (Fig. 80). That the deposit was made in ancient times on land and not in the sea is proved by the remains which it contains of land animals and plants of species now extinct. That it was laid by rivers and not by fresh-water lakes is shown by its structure. Wide stretches of flat-lying, clays and sands are interrupted by long, narrow belts of gravel which mark the channels of the ancient streams. Gravels, and sands are often cross bedded, and their well worn pebbles may be identified with the rocks of the mountains. After building this sheet of waste the streams ceased to aggrade and began the work of destruction. Large uneroded remnants, their surfaces flat as a floor, remain as the High Plains of western Kansas and Nebraska.

RIVER DEPOSITS IN SUBSIDING TROUGHS. To a geologist the most important river deposits are those which gather in areas of gradual subsidence; they are often of vast extent and immense thickness, and such deposits of past geological ages have not infrequently been preserved, with all their records of the times in which they were built, by being carried below the level of the sea, to be brought to light by a later uplift. On the other hand, river deposits which remain above baselevels of erosion are swept away comparatively soon.

THE GREAT VALLEY OF CALIFORNIA is a monotonously level plain of great fertility, four hundred miles in length and fifty miles in average width, built of waste swept down by streams from the mountain ranges which inclose it,--the Sierra Nevada on the east and the Coast Range on the west. On the waste slopes at the foot of the bordering hills coarse gravels and even bowlders are left, while over the interior the slow-flowing streams at times of flood spread wide sheets of silt. Organic deposits are now forming by the decay of vegetation in swampy tule (reed) lands and in shallow lakes which occupy depressions left by the aggrading streams.

Deep borings show that this great trough is filled to a depth of at least two thousand feet below sea level with recent unconsolidated sands and silts containing logs of wood and fresh- water shells. These are land deposits, and the absence of any marine deposits among them proves that the region has not been invaded by the sea since the accumulation began. It has therefore been slowly subsiding and its streams, although continually carried below grade, have yet been able to aggrade the surface as rapidly as the region sank, and have maintained it, as at present, slightly above sea level.

THE INDO-GANGETIC PLAIN, spread by the Brahmaputra, the Ganges, and the Indus river systems, stretches for sixteen hundred miles along the southern base of the Himalaya Mountains and occupies an area of three hundred thousand square miles (Fig.342). It consists of the flood plains of the master streams and the confluent fans of the tributaries which issue from the mountains on the north. Large areas are subject to overflow each season of flood, and still larger tracts mark abandoned flood plains below which the rivers have now cut their beds. The plain is built of far- stretching beds of clay, penetrated by streaks of sand, and also of gravel near the mountains. Beds of impure peat occur in it, and it contains fresh-water shells and the bones of land animals of species now living in northern India. At Lucknow an artesian well was sunk to one thousand feet below sea level without reaching the bottom of these river-laid sands and silts, proving a slow subsidence with which the aggrading rivers have kept pace.

WARPED VALLEYS. It is not necessary that an area should sink below sea level in order to be filled with stream-swept waste. High valleys among growing mountain ranges may suffer warping, or may be blockaded by rising mountain folds athwart them. Where the deformation is rapid enough, the river may be ponded and the valley filled with lake-laid sediments. Even when the river is able to maintain its right of way it may yet have its declivity so lessened that it is compelled to aggrade its course continually, filling the valley with river deposits which may grow to an enormous thickness.

Behind the outer ranges of the Himalaya Mountains lie several waste-filled valleys, the largest of which are Kashmir and Nepal, the former being an alluvial plain about as large as the state of Delaware. The rivers which drain these plains have already cut down their outlet gorges sufficiently to begin the task of the removal of the broad accumulations which they have brought in from the surrounding mountains. Their present flood plains lie as much as some hundreds of feet below wide alluvial terraces which mark their former levels. Indeed, the horizontal beds of the Hundes Valley have been trenched to the depth of nearly three thousand feet by the Sutlej River. These deposits are recent or subrecent, for there have been found at various levels the remains of land plants and land and fresh-water shells, and in some the bones of such animals as the hyena and the goat, of species or of genera now living. Such soft deposits cannot be expected to endure through any considerable length of future time the rapid erosion to which their great height above the level of the sea will subject them.

CHARACTERISTICS OF RIVER DEPOSITS. The examples just cited teach clearly the characteristic features of extensive river deposits. These deposits consist of broad, flat-lying sheets of clay and fine sand left by the overflow at time of flood, and traversed here and there by long, narrow strips of coarse, cross-bedded sands and gravels thrown down by the swifter currents of the shifting channels. Occasional beds of muck mark the sites of shallow lakelets or fresh-water swamps. The various strata also contain some remains of the countless myriads of animals and plants which live upon the surface of the plain as it is in process of building. River shells such as the mussel, land shells such as those of snails, the bones of fishes and of such land animals as suffer drowning at times of flood or are mired in swampy places, logs of wood, and the stems and leaves of plants are examples of the variety of the remains of land and fresh-water organisms which are entombed in river deposits and sealed away as a record of the life of the time, and as proof that the deposits were laid by streams and not beneath the sea.

BASIN DEPOSITS

DEPOSITS IN DRY BASINS. On desert areas without outlet to the sea, as on the Great Basin of the United States and the deserts of central Asia, stream-swept waste accumulates indefinitely. The rivers of the surrounding mountains, fed by the rains and melting snows of these comparatively moist elevations, dry and soak away as they come down upon the arid plains. They are compelled to lay aside their entire load of waste eroded from the mountain valleys, in fans which grow to enormous size, reaching in some instances thousands of feet in thickness.

The monotonous levels of Turkestan include vast alluvial tracts now in process of building by the floods of the frequently shifting channels of the Oxus and other rivers of the region. For about seven hundred miles from its mouth in Aral Lake the Oxus receives no tributaries, since even the larger branches of its system are lost in a network of distributaries and choked with desert sands before they reach their master stream. These aggrading rivers, which have channels but no valleys, spread their muddy floods--which in the case of the Oxus sometimes equal the average volume of the Mississippi--far and wide over the plain, washing the bases of the desert dunes.

PLAYAS. In arid interior basins the central depressions may be occupied by playas,--plains of fine mud washed forward from the margins. In the wet season the playa is covered with a thin sheet of muddy water, a playa lake, supplied usually by some stream at flood. In the dry season the lake evaporates, the river which fed it retreats, and there is left to view a hard, smooth, level floor of sun-baked and sun-cracked yellow clay utterly devoid of vegetation.

In the Black Rock desert of Nevada a playa lake spreads over an area fifty miles long and twenty miles wide. In summer it disappears; the Quinn River, which feeds it, shrinks back one hundred miles toward its source, leaving an absolutely barren floor of clay, level as the sea.

LAKE DEPOSITS. Regarding lakes as parts of river systems, we may now notice the characteristic features of the deposits in lake basins. Soundings in lakes of considerable size and depth show that their bottoms are being covered with tine clays. Sand and gravel are found along; their margins, being brought in by streams and worn by waves from the shore, but there are no tidal or other strong currents to sweep coarse waste out from shore to any considerable distance. Where fine clays are now found on the land in even, horizontal layers containing the remains of fresh-water animals and plants, uncut by channels tilled with cross-bedded gravels and sands and bordered by beach deposits of coarse waste, we may safely infer the existence of ancient lakes.

MARL. Marl is a soft, whitish deposit of carbonate of lime, mingled often with more or less of clay, accumulated in small lakes whose feeding springs are charged with carbonate of lime and into which little waste is washed from the land. Such lakelets are not infrequent on the surface of the younger drift sheets of Michigan and northern Indiana, where their beds of marl--sometimes as much as forty feet thick--are utilized in the manufacture of Portland cement. The deposit results from the decay of certain aquatic plants which secrete lime carbonate from the water, from the decomposition of the calcareous shells of tiny mollusks which live in countless numbers on the lake floor, and in some cases apparently from chemical precipitation.

PEAT. We have seen how lakelets are extinguished by the decaying remains of the vegetation which they support. A section of such a fossil lake shows that below the growing mosses and other plants of the surface of the bog lies a spongy mass composed of dead vegetable tissue, which passes downward gradually into PEAT,--a dense, dark brown carbonaceous deposit in which, to the unaided eye, little or no trace of vegetable structure remains. When dried, peat forms a fuel of some value and is used either cut into slabs and dried or pressed into bricks by machinery.

When vegetation decays in open air the carbon of its tissues, taken from the atmosphere by the leaves, is oxidized and returned to it in its original form of carbon dioxide. But decomposing in the presence of water, as in a bog, where the oxygen of the air is excluded, the carbonaceous matter of plants accumulates in deposits of peat.

Peat bogs are numerous in regions lately abandoned by glacier ice, where river systems are so immature that the initial depressions left in the sheet of drift spread over the country have not yet been drained. One tenth of the surface of Ireland is said to be covered with peat, and small bogs abound in the drift-covered area of New England and the states lying as far west as the Missouri River. In Massachusetts alone it has been reckoned that there are fifteen billion cubic feet of peat, the largest bog occupying several thousand acres.

Much larger swamps occur on the young coastal plain of the Atlantic from New Jersey to Florida. The Dismal Swamp, for example, in Virginia and North Carolina is forty miles across. It is covered with a dense growth of water-loving trees such as the cypress and black gum. The center of the swamp is occupied by Lake Drummond, a shallow lake seven miles in diameter, with banks of pure-peat, and still narrowing from the encroachment of vegetation along its borders.

SALT LAKES. In arid climates a lake rarely receives sufficient inflow to enable it to rise to the basin rim and find an outlet. Before this height is reached its surface becomes large enough to discharge by evaporation into the dry air the amount of water that is supplied by streams. As such a lake has no outlet, the minerals in solution brought into it by its streams cannot escape from the basin. The lake water becomes more and more heavily charged with such substances as common salt and the sulphates and carbonates of lime, of soda, and of potash, and these are thrown down from solution one after another as the point of saturation for each mineral is reached. Carbonate of lime, the least soluble and often the most abundant mineral brought in, is the first to be precipitated. As concentration goes on, gypsum, which is insoluble in a strong brine, is deposited, and afterwards common salt. As the saltness of the lake varies with the seasons and with climatic changes, gypsum and salt are laid in alternate beds and are interleaved with sedimentary clays spread from the waste brought in by streams at times of flood. Few forms of life can live in bodies of salt water so concentrated that chemical deposits take place, and hence the beds of salt, gypsum, and silt of such lakes are quite barren of the remains of life. Similar deposits are precipitated by the concentration of sea water in lagoons and arms of the sea cut off from the ocean.

LAKES BONNEVILLE AND LAHONTAN. These names are given to extinct lakes which once occupied large areas in the Great Basin, the former in Utah, the latter in northwestern Nevada. Their records remain in old horizontal beach lines which they drew along their mountainous shores at the different levels at which they stood, and in the deposits of their beds. At its highest stage Lake Bonneville, then one thousand feet deep, overflowed to the north and was a fresh-water lake. As it shrank below the outlet it became more and more salty, and the Great Salt Lake, its withered residue, is now depositing salt along its shores. In its strong brine lime carbonate is insoluble, and that brought in by streams is thrown down at once in the form of travertine.

Lake Lahontan never had an outlet. The first chemical deposits to be made along its shores were deposits of travertine, in places eighty feet thick. Its floor is spread with fine clays, which must have been laid in deep, still water, and which are charged with the salts absorbed by them as the briny water of the lake dried away. These sedimentary clays are in two divisions, the upper and lower, each being about one hundred feet thick. They are separated by heavy deposits of well-rounded, cross-bedded gravels and sands, similar to those spread at the present time by the intermittent streams of arid regions. A similar record is shown in the old floors of Lake Bonneville. What conclusions do you draw from these facts as to the history of these ancient lakes?

DELTAS

In the river deposits which are left above sea level particles of waste are allowed to linger only for a time. From alluvial fans and flood plains they are constantly being taken up and swept farther on downstream. Although these land forms may long persist, the particles which compose them are ever changing. We may therefore think of the alluvial deposits of a valley as a stream of waste fed by the waste mantle as it creeps and washes down the valley sides, and slowly moving onwards to the sea.

In basins waste finds a longer rest, but sooner or later lakes and dry basins are drained or filled, and their deposits, if above sea level, resume their journey to their final goal. It is only when carried below the level of the sea that they are indefinitely preserved.

On reaching this terminus, rivers deliver their load to the ocean. In some cases the ocean is able to take it up by means of strong tidal and other currents, and to dispose of it in ways which we shall study later. But often the load is so large, or the tides are so weak, that much of the waste which the river brings in settles at its mouth, there building up a deposit called the DELTA, from the Greek letter of that name, whose shape it sometimes resembles.

Deltas and alluvial fans have many common characteristics. Both owe their origin to a sudden check in the velocity of the river, compelling a deposit of the load; both are triangular in outline, the apex pointing upstream; and both are traversed by distributaries which build up all parts in turn.

In a delta we may distinguish deposits of two distinct kinds,-- the submarine and the subaerial. In part a delta is built of waste brought down by the river and redistributed and spread by waves and tides over the sea bottom adjacent to the river's mouth. The origin of these deposits is recorded in the remains of marine animals and plants which they contain.

As the submarine delta grows near to the level of the sea the distributaries of the river cover it with subaerial deposits altogether similar to those of the flood plain, of which indeed the subaerial delta is the prolongation. Here extended deposits of peat may accumulate in swamps, and the remains of land and fresh- water animals and plants swept down by the stream are imbedded in the silts laid at times of flood.

Borings made in the deltas of great rivers such as the Mississippi, the Ganges, and the Nile, show that the subaerial portion often reaches a surprising thickness. Layers of peat, old soils, and forest grounds with the stumps of trees are discovered hundreds of feet below sea level. In the Nile delta some eight layers of coarse gravel were found interbedded with river silts, and in the Ganges delta at Calcutta a boring nearly five hundred feet in depth stopped in such a layer.

The Mississippi has built a delta of twelve thousand three hundred square miles, and is pushing the natural embankments of its chief distributaries into the Gulf at a maximum rate of a mile in sixteen years. Muddy shoals surround its front, shallow lakes, e.g. lakes Pontchartrain and Borgne, are formed between the growing delta and the old shore line, and elongate lakes and swamps are inclosed between the natural embankments of the distributaries.

The delta of the Indus River, India, lies so low along shore that a broad tract of country is overflowed by the highest tides. The submarine portion of the delta has been built to near sea level over so wide a belt offshore that in many places large vessels cannot come even within sight of land because of the shallow water.

A former arm of the sea, the Rann of Cutch, adjoining the delta on the east has been silted up and is now an immense barren flat of sandy mud two hundred miles in length and one hundred miles in greatest breadth. Each summer it is flooded with salt water when the sea is brought in by strong southwesterly monsoon winds, and the climate during the remainder of the year is hot and dry. By the evaporation of sea water the soil is thus left so salty that no vegetation can grow upon it, and in places beds of salt several feet in thickness have accumulated. Under like conditions salt beds of great thickness have been formed in the past and are now found buried among the deposits of ancient deltas.

SUBSIDENCE OF GREAT DELTAS. As a rule great deltas are slowly sinking. In some instances upbuilding by river deposits has gone on as rapidly as the region has subsided. The entire thickness of the Ganges delta, for example, so far as it has been sounded, consists of deposits laid in open air. In other cases interbedded limestones and other sedimentary rocks containing marine fossils prove that at times subsidence has gained on the upbuilding and the delta has been covered with the sea.

It is by gradual depression that delta deposits attain enormous thickness, and, being lowered beneath the level of the sea, are safely preserved from erosion until a movement of the earth's crust in the opposite direction lifts them to form part of the land. We shall read later in the hard rocks of our continent the records of such ancient deltas, and we shall not be surprised to find them as thick as are those now building at the mouths of great rivers.

LAKE DELTAS. Deltas are also formed where streams lose their velocity on entering the still waters of lakes. The shore lines of extinct lakes, such as Lake Agassiz and Lakes Bonneville and Lahontan, may be traced by the heavy deposits at the mouths of their tributary streams.

We have seen that the work of streams is to drain the lands of the water poured upon them by the rainfall, to wear them down, and to carry their waste away to the sea, there to be rebuilt by other agents into sedimentary rocks. The ancient strata of which the continents are largely made are composed chiefly of material thus worn from still more ancient lands--lands with their hills and valleys like those of to-day--and carried by their rivers to the ocean. In all geological times, as at the present, the work of streams has been to destroy the lands, and in so doing to furnish to the ocean the materials from which the lands of future ages were to be made. Before we consider how the waste of the land brought in by streams is rebuilt upon the ocean floor, we must proceed to study the work of two agents, glacier ice and the wind, which cooperate with rivers in the denudation of the land.

CHAPTER V

THE WORK OF GLACIERS

THE DRIFT. The surface of northeastern North America, as far south as the Ohio and Missouri rivers, is generally covered by the drift,--a formation which is quite unlike any which we have so far studied. A section of it, such as that illustrated in Figure 87, shows that for the most part it is unstratified, consisting of clay, sand, pebbles, and even large bowlders, all mingled pell-mell together. The agent which laid the drift is one which can carry a load of material of all sizes, from the largest bowlder to the finest clay, and deposit it without sorting.

The stones of the drift are of many kinds. The region from which it was gathered may well have been large in order to supply these many different varieties of rocks. Pebbles and bowlders have been left far from their original homes, as may be seen in southern Iowa, where the drift contains nuggets of copper brought from the region about Lake Superior. The agent which laid the drift is one able to gather its load over a large area and carry it a long way.

The pebbles of the drift are unlike those rounded by running water or by waves. They are marked with scratches. Some are angular, many have had their edges blunted, while others have been ground flat and smooth on one or more sides, like gems which have been faceted by being held firmly against the lapidary's wheel. In many places the upper surface of the country rock beneath the drift has been swept clean of residual clays and other waste. All rock rotten has been planed away, and the ledges of sound rock to which the surface has been cut down have been rubbed smooth and scratched with long, straight, parallel lines. The agent which laid the drift can hold sand and pebbles firmly in its grasp and can grind them against the rock beneath, thus planing it down and scoring it, while faceting the pebbles also.

Neither water nor wind can do these things. Indeed, nothing like the drift is being formed by any process now at work anywhere in the eastern United States. To find the agent which has laid this extensive formation we must go to a region of different climatic conditions.

THE INLAND ICE OF GREENLAND. Greenland is about fifteen hundred miles long and nearly seven hundred miles in greatest width. With the exception of a narrow fringe of mountainous coast land, it is completely buried beneath a sheet of ice, in shape like a vast white shield, whose convex surface rises to a height of nine thousand feet above the sea. The few explorers who have crossed the ice cap found it a trackless desert destitute of all life save such lowly forms as the microscopic plant which produces the so-called "red snow." On the smooth plain of the interior no rock waste relieves the snow's dazzling whiteness; no streams of running water are seen; the silence is broken only by howling storm winds and the rustle of the surface snow which they drive before them. Sounding with long poles, explorers find that below the powdery snow of the latest snowfall lie successive layers of earlier snows, which grow more and more compact downward, and at last have altered to impenetrable ice. The ice cap formed by the accumulated snows of uncounted centuries may well be more than a mile in depth. Ice thus formed by the compacting of snow is distinguished when in motion as GLACIER ICE.

The inland ice of Greenland moves. It flows with imperceptible slowness under its own weight, like, a mass of some viscous or plastic substance, such as pitch or molasses candy, in all directions outward toward the sea. Near the edge it has so thinned that mountain peaks are laid bare, these islands in the sea of ice being known as NUNATAKS. Down the valleys of the coastal belt it drains in separate streams of ice, or GLACIERS. The largest of these reach the sea at the head of inlets, and are therefore called TIDE GLACIERS. Their fronts stand so deep in sea water that there is visible seldom more than three hundred feet of the wall of ice, which in many glaciers must be two thousand and more feet high. From the sea walls of tide glaciers great fragments break off and float away as icebergs. Thus snows which fell in the interior of this northern land, perhaps many thousands of years ago, are carried in the form of icebergs to melt at last in the North Atlantic.

Greenland, then, is being modeled over the vast extent of its interior not by streams of running water, as are regions in warm and humid climates, nor by currents of air, as are deserts to a large extent, but by a sheet of flowing ice. What the ice sheet is doing in the interior we may infer from a study of the separate glaciers into which it breaks at its edge.

THE SMALLER GREENLAND GLACIERS. Many of the smaller glaciers of Greenland do not reach the sea, but deploy on plains of sand and gravel. The edges of these ice tongues are often as abrupt as if sliced away with a knife (Fig. 92), and their structure is thus readily seen. They are stratified, their layers representing in part the successive snowfalls of the interior of the country. The upper layers are commonly white and free from stones; but the lower layers, to the height of a hundred feet or more, are dark with debris which is being slowly carried on. So thickly studded with stones is the base of the ice that it is sometimes difficult to distinguish it from the rock waste which has been slowly dragged beneath the glacier or left about its edges. The waste beneath and about the glacier is unsorted. The stones are of many kinds, and numbers of them have been ground to flat faces. Where the front of the ice has retreated the rock surface is seen to be planed and scored in places by the stones frozen fast in the sole of the glacier.

We have now found in glacier ice an agent able to produce the drift of North America. The ice sheet of Greenland is now doing what we have seen was done in the recent past in our own land. It is carrying for long distances rocks of many kinds gathered, we may infer, over a large extent of country. It is laying down its load without assortment in unstratified deposits. It grinds down and scores the rock over which it moves, and in the process many of the pebbles of its load are themselves also ground smooth and scratched. Since this work can be done by no other agent, we must conclude that the northeastern part of our own continent was covered in the recent past by glacier ice, as Greenland is to-day.

VALLEY GLACIERS

The work of glacier ice can be most conveniently studied in the separate ice streams which creep down mountain valleys in many regions such as Alaska, the western mountains of the United States and Canada, the Himalayas, and the Alps. As the glaciers of the Alps have been studied longer and more thoroughly than any others, we shall describe them in some detail as examples of valley glaciers in all parts of the world.

CONDITIONS OF GLACIER FORMATION. The condition of the great accumulation of snow to which glaciers are due--that more or less of each winter's snow should be left over unmelted and unevaporated to the next--is fully met in the Alps. There is abundant moisture brought by the winds from neighboring seas. The currents of moist air driven up the mountain slopes are cooled by their own expansion as they rise, and the moisture which they contain is condensed at a temperature at or below 32 degrees F., and therefore is precipitated in the form of snow. The summers are cool and their heat does not suffice to completely melt the heavy snow of the preceding winter. On the Alps the SNOW LINE--the lower limit of permanent snow--is drawn at about eight thousand five hundred feet above sea level. Above the snow line on the slopes and crests, where these are not too steep, the snow lies the year round and gathers in valley heads to a depth of hundreds of feet.

This is but a small fraction of the thickness to which snow would be piled on the Alps were it not constantly being drained away. Below the snow fields which mantle the heights the mountain valleys are occupied by glaciers which extend as much as a vertical mile below the snow line. The presence in the midst of forests and meadows and cultivated fields of these tongues of ice, ever melting and yet from year to year losing none of their bulk, proves that their loss is made good in the only possible way. They are fed by snow fields above, whose surplus of snow they drain away in the form of ice. The presence of glaciers below the snow line is a clear proof that, rigid and motionless as they appear, glaciers really are in constant motion down valley.

THE NEVE FIELD. The head of an Alpine valley occupied by a glacier is commonly a broad amphitheater deeply filled with snow. Great peaks tower above it, and snowy slopes rise on either side on the flanks of mountain spurs. From these heights fierce winds drift the snows into the amphitheater, and avalanches pour in their torrents of snow and waste. The snow of the amphitheater is like that of drifts in late winter after many successive thaws and freezings. It is made of hard grains and pellets and is called NEVE. Beneath the surface of the neve field and at its outlet the granular neve has been compacted to a mass of porous crystalline ice. Snow has been changed to neve, and neve to glacial ice, both by pressure, which drives the air from the interspaces of the snowflakes, and also by successive meltings and freezings, much as a snowball is packed in the warm hand and becomes frozen to a ball of ice.

THE BERGSCHRUND. The neve is in slow motion. It breaks itself loose from the thinner snows about it, too shallow to share its motion, and from the rock rim which surrounds it, forming a deep fissure called the bergschrund, sometimes a score and more feet wide.

SIZE OF GLACIERS. The ice streams of the Alps vary in size according to the amount of precipitation and the area of the neve fields which they drain.

The largest of Alpine glaciers, the Aletsch, is nearly ten miles long and has an average width of about a mile. The thickness of some of the glaciers of the Alps is as much as a thousand feet. Giant glaciers more than twice the length of the longest in the Alps occur on the south slope of the Himalaya Mountains, which receive frequent precipitations of snow from moist winds from the Indian Ocean. The best known of the many immense glaciers of Alaska, the Muir, has an area of about eight hundred square miles (Fig. 95).

GLACIER MOTION. The motion of the glaciers of the Alps seldom exceeds one or two feet a day. Large glaciers, because of the enormous pressure of their weight and because of less marginal resistance, move faster than small ones. The Muir advances at the rate of seven feet a day, and some of the larger tide glaciers of Greenland are reported to move at the exceptional rate of fifty feet and more in the same time. Glaciers move faster by day than by night, and in summer than in winter. Other laws of glacier motion may be discovered by a study of Figures 96 and 97. It is important to remember that glaciers do not slide bodily over their beds, but urged by gravity move slowly down valley in somewhat the same way as would a stream of thick mud. Although small pieces of ice are brittle, the large mass of granular ice which composes a glacier acts as a viscous substance.

CREVASSES. Slight changes of slope in the glacier bed, and the different rates of motion in different parts, produce tensions under which the ice cracks and opens in great fissures called crevasses. At an abrupt descent in the bed the ice is shattered into great fragments, which unite again below the icefall. Crevasses are opened on lines at right angles to the direction of the tension. TRANSVERSE CREVASSES are due to a convexity in the bed which stretches the ice lengthwise (Fig. 99). MARGINAL CREVASSES are directed upstream and inwards; RADIAL CREVASSES are found where the ice stream deploys from some narrow valley and spreads upon some more open space. What is the direction of the tension which causes each and to what is it due?

LATERAL AND MEDIAL MORAINES. The surface of a glacier is striped lengthwise by long dark bands of rock debris. Those in the center are called the medial moraines. The one on either margin is a lateral moraine, and is clearly formed of waste which has fallen on the edge of the ice from the valley slopes. A medial moraine cannot be formed in this way, since no rock fragments can fall so far out from the sides. But following it up the glacial stream, one finds that a medial moraine takes its beginning at the junction of the glacier and some tributary and is formed by the union of their two adjacent lateral moraines. Each branch thus adds a medial moraine, and by counting the number of medial moraines of a trunk stream one may learn of how many branches it is composed.

Surface moraines appear in the lower course of the glacier as ridges, which may reach the exceptional height of one hundred feet. The bulk of such a ridge is ice. It has been protected from the sun by the veneer of moraine stuff; while the glacier surface on either side has melted down at least the distance of the height of the ridge. In summer the lowering of the glacial surface by melting goes on rapidly. In Swiss glaciers it has been estimated that the average lowering of the surface by melting and evaporation amounts to ten feet a year. As a moraine ridge grows higher and more steep by the lowering of the surface of the surrounding ice, the stones of its cover tend to slip down its sides. Thus moraines broaden, until near the terminus of a glacier they may coalesce in a wide field of stony waste.

ENGLACIAL DRIFT. This name is applied to whatever debris is carried within the glacier. It consists of rock waste fallen on the neve and there buried by accumulations of snow, and of that engulfed in the glacier where crevasses have opened beneath a surface moraine. As the surface of the glacier is lowered by melting, more or less englacial drift is brought again to open air, and near the terminus it may help to bury the ice from view beneath a sheet of debris.

THE GROUND MORAINE. The drift dragged along at the glacier's base and lodged beneath it is known as the ground moraine. Part of the material of it has fallen down deep crevasses and part has been torn and worn from the glacier's bed and banks. While the stones of the surface moraines remain as angular as when they lodged on the ice, many of those of the ground moraine have been blunted on the edges and faceted and scratched by being ground against one another and the rocky bed.

In glaciers such as those of Greenland, whose basal layers are well loaded with drift and whose surface layers are nearly clean, different layers have different rates of motion, according to the amount of drift with which they are clogged. One layer glides over another, and the stones inset in each are ground and smoothed and scratched. Usually the sides of glaciated pebbles are more worn than the ends, and the scratches upon them run with the longer axis of the stone. Why?

THE TERMINAL MORAINE. As a glacier is in constant motion, it brings to its end all of its load except such parts of the ground moraine as may find permanent lodgment beneath the ice. Where the glacier front remains for some time at one place, there is formed an accumulation of drift known as the terminal moraine. In valley glaciers it is shaped by the ice front to a crescent whose convex side is downstream. Some of the pebbles of the terminal moraine are angular, and some are faceted and scored, the latter having come by the hard road of the ground moraine. The material of the dump is for the most part unsorted, though the water of the melting ice may find opportunity to leave patches of stratified sands and gravels in the midst of the unstratified mass of drift, and the finer material is in places washed away.

GLACIER DRAINAGE. The terminal moraine is commonly breached by a considerable stream, which issues from beneath the ice by a tunnel whose portal has been enlarged to a beautiful archway by melting in the sun and the warm air (Fig. 107). The stream is gray with silt and loaded with sand and gravel washed from the ground moraine. "Glacier milk" the Swiss call this muddy water, the gray color of whose silt proves it rock flour freshly ground by the ice from the unoxidized sound rock of its bed, the mud of streams being yellowish when it is washed from the oxidized mantle of waste. Since glacial streams are well loaded with waste due to vigorous ice erosion, the valley in front of the glacier is commonly aggraded to a broad, flat floor. These outwash deposits are known as VALLEY DRIFT.

The sand brought out by streams from beneath a glacier differs from river sand in that it consists of freshly broken angular grains. Why?

The stream derives its water chiefly from the surface melting of the glacier. As the ice is touched by the rays of the morning sun in summer, water gathers in pools, and rills trickle and unite in brooklets which melt and cut shallow channels in the blue ice. The course of these streams is short. Soon they plunge into deep wells cut by their whirling waters where some crevasse has begun to open across their path. These wells lead into chambers and tunnels by which sooner or later their waters find way to the rock floor of the valley and there unite in a subglacial stream.

THE LOWER LIMIT OF GLACIERS. The glaciers of a region do not by any means end at a uniform height above sea level. Each terminates where its supply is balanced by melting. Those therefore which are fed by the largest and deepest neves and those also which are best protected from the sun by a northward exposure or by the depth of their inclosing valleys flow to lower levels than those whose supply is less and whose exposure to the sun is greater.

A series of cold, moist years, with an abundant snowfall, causes glaciers to thicken and advance; a series of warm, dry years causes them to wither and melt back. The variation in glaciers is now carefully observed in many parts of the world. The Muir glacier has retreated two miles in twenty years. The glaciers of the Swiss Alps are now for the most part melting back, although a well-known glacier of the eastern Alps, the Vernagt, advanced five hundred feet in the year 1900, and was then plowing up its terminal moraine.

How soon would you expect a glacier to advance after its neve fields have been swollen with unusually heavy snows, as compared with the time needed for the flood of a large river to reach its mouth after heavy rains upon its headwaters?

On the surface of glaciers in summer time one may often see large stones supported by pillars of ice several feet in height (Fig. 108). These "glacier tables" commonly slope more or less strongly to the south, and thus may be used to indicate roughly the points of the compass. Can you explain their formation and the direction of their slope? On the other hand, a small and thin stone, or a patch of dust, lying on the ice, tends to sink a few inches into it. Why?

In what respects is a valley glacier like a mountain stream which flows out upon desert plains?

Two confluent glaciers do not mingle their currents as do two confluent rivers. What characteristics of surface moraines prove this fact?

What effect would you expect the laws of glacier motion to have on the slant of the sides of transverse crevasses?

A trunk glacier has four medial moraines. Of how many tributaries is it composed? Illustrate by diagram.

State all the evidences which you have found that glaciers move.

If a glacier melts back with occasional pauses up a valley, what records are left of its retreat?

PIEDMONT GLACIERS

THE MALASPINA GLACIER. Piedmont (foot of the mountain) glaciers are, as the name implies, ice fields formed at the foot of mountains by the confluence of valley glaciers. The Malaspina glacier of Alaska, the typical glacier of this kind, is seventy miles wide and stretches for thirty miles from the foot of the Mount Saint Elias range to the shore of the Pacific Ocean. The valley glaciers which unite and spread to form this lake of ice lie above the snow line and their moraines are concealed beneath neve. The central area of the Malaspina is also free from debris; but on the outer edge large quantities of englacial drift are exposed by surface melting and form a belt of morainic waste a few feet thick and several miles wide, covered in part with a luxuriant forest, beneath which the ice is in places one thousand feet in depth. The glacier here is practically stagnant, and lakes a few hundred yards across, which could not exist were the ice in motion and broken with crevasses, gather on their beds sorted waste from the moraine. The streams which drain the glacier have cut their courses in englacial and subglacial tunnels; none flow for any distance on the surface. The largest, the Yahtse River, issues from a high archway in the ice,--a muddy torrent one hundred feet wide and twenty feet deep, loaded with sand and stones which it deposits in a broad outwash plain (Fig. 110). Where the ice has retreated from the sea there is left a hummocky drift sheet with hollows filled with lakelets. These deposits help to explain similar hummocky regions of drift and similar plains of coarse, water-laid material often found in the drift-covered area of the northeastern United States.

THE GEOLOGICAL WORK OF GLACIER ICE

The sluggish glacier must do its work in a different way from the agile river. The mountain stream is swift and small, and its channel occupies but a small portion of the valley. The glacier is slow and big; its rate of motion may be less than a millionth of that of running water over the same declivity, and its bulk is proportionately large and fills the valley to great depth. Moreover, glacier ice is a solid body plastic under slowly applied stresses, while the water of rivers is a nimble fluid.

TRANSPORTATION. Valley glaciers differ from rivers as carriers in that they float the major part of their load upon their surface, transporting the heaviest bowlder as easily as a grain of sand; while streams push and roll much of their load along their beds, and their power of transporting waste depends solely upon their velocity. The amount of the surface load of glaciers is limited only by the amount of waste received from the mountain slopes above them. The moving floor of ice stretched high across a valley sweeps along as lateral moraines much of the waste which a mountain stream would let accumulate in talus and alluvial cones.

While a valley glacier carries much of its load on top, an ice sheet, such as that of Greenland, is free from surface debris, except where moraines trail away from some nunatak. If at its edge it breaks into separate glaciers which drain down mountain valleys, these tongues of ice will carry the selvages of waste common to valley glaciers. Both ice sheets and valley glaciers drag on large quantities of rock waste in their ground moraines.

Stones transported by glaciers are sometimes called erratics. Such are the bowlders of the drift of our northern states. Erratics may be set down in an insecure position on the melting of the ice.

DEPOSIT. Little need be added here to what has already been said of ground and terminal moraines. All strictly glacial deposits are unstratified. The load laid down at the end of a glacier in the terminal moraine is loose in texture, while the drift lodged beneath the glacier as ground moraine is often an extremely dense, stony clay, having been compacted under the pressure of the overriding ice.

EROSION. A glacier erodes its bed and banks in two ways,--by abrasion and by plucking.

The rock bed over which a glacier has moved is seen in places to have been abraded, or ground away, to smooth surfaces which are marked by long, straight, parallel scorings aligned with the line of movement of the ice and varying in size from hair lines and coarse scratches to exceptional furrows several feet deep. Clearly this work has been accomplished by means of the sharp sand, the pebbles, and the larger stones with which the base of the glacier is inset, and which it holds in a firm grasp as running water cannot. Hard and fine-grained rocks, such as granite and quartzite, are often not only ground down to a smooth surface but are also highly polished by means of fine rock flour worn from the glacier bed.

In other places the bed of the glacier is rough and torn. The rocks have been disrupted and their fragments have been carried away,--a process known as PLUCKING. Moving under immense pressure the ice shatters the rock, breaks off projections, presses into crevices and wedges the rocks apart, dislodges the blocks into which the rock is divided by joints and bedding planes, and freezing fast to the fragments drags them on. In this work the freezing and thawing of subglacial waters in any cracks and crevices of the rock no doubt play an important part. Plucking occurs especially where the bed rock is weak because of close jointing. The product of plucking is bowlders, while the product of abrasion is fine rock flour and sand.

Is the ground moraine of Figure 87 due chiefly to abrasion or to plucking?

ROCHES MOUTONNEES AND ROUNDED HILLS. The prominences left between the hollows due to plucking are commonly ground down and rounded on the stoss side,--the side from which the ice advances,--and sometimes on the opposite, the lee side, as well. In this way the bed rock often comes to have a billowy surface known as roches moutonnees (sheep rocks). Hills overridden by an ice sheet often have similarly rounded contours on the stoss side, while on the lee side they may be craggy, either because of plucking or because here they have been less worn from their initial profile.

THE DIRECTION OF GLACIER MOVEMENT. The direction of the flow of vanished glaciers and ice sheets is recorded both in the differences just mentioned in the profiles of overridden hills and also in the minute details of the glacier trail.

Flint nodules or other small prominences in the bed rock are found more worn on the stoss than on the lee side, where indeed they may have a low cone of rock protected by them from abrasion. Cavities, on the other hand, have their edges worn on the lee side and left sharp upon the stoss.

Surfaces worn and torn in the ways which we have mentioned are said to be glaciated. But it must not be supposed that a glacier everywhere glaciates its bed. Although in places it acts as a rasp or as a pick, in others, and especially where its pressure is least, as near the terminus, it moves over its bed in the manner of a sled. Instances are known where glaciers have advanced over deposits of sand and gravel without disturbing them to any notable degree. Like a river, a glacier does not everywhere erode. In places it leaves its bed undisturbed and in places aggrades it by deposits of the ground moraine.

CIRQUES. Valley glaciers commonly head as we have seen, in broad amphitheaters deeply filled with snow and ice. On mountains now destitute of glaciers, but whose glaciation shows that they have supported glaciers in the past, there are found similar crescentic hollows with high, precipitous walls and glaciated floors. Their floors are often basined and hold lakelets whose deep and quiet waters reflect the sheltering ramparts of rugged rock which tower far above them. Such mountain hollows are termed CIRQUES. As a powerful spring wears back a recess in the valley side where it discharges, so the fountain head of a glacier gradually wears back a cirque. In its slow movement the neve field broadly scours its bed to a flat or basined floor. Meanwhile the sides of the valley head are steepened and driven back to precipitous walls. For in winter the crevasse of the bergschrund which surrounds the neve field is filled with snow and the neve is frozen fast to the rocky sides of the valley. In early summer the neve tears itself free, dislodging and removing any loosened blocks, and the open fissure of the bergschrund allows frost and other agencies of weathering to attack the unprotected rock. As cirques are thus formed and enlarged the peaks beneath which they lie are sharpened, and the mountain crests are scalloped and cut back from either side to knife-edged ridges.

In the western mountains of the United States many cirques, now empty of neve and glacier ice, and known locally as "basins," testify to the fact that in recent times the snow line stood beneath the levels of their floors, and thus far below its present altitude.

GLACIER TROUGHS. The channel worn to accommodate the big and clumsy glacier differs markedly from the river valley cut as with a saw by the narrow and flexible stream and widened by the weather and the wash of rains. The valley glacier may easily be from one thousand to three thousand feet deep and from one to three miles wide. Such a ponderous bulk of slowly moving ice does not readily adapt itself to sharp turns and a narrow bed. By scouring and plucking all resisting edges it develops a fitting channel with a wide, flat floor, and steep, smooth sides, above which are seen the weathered slopes of stream-worn mountain valleys. Since the trunk glacier requires a deeper channel than do its branches, the bed of a branch glacier enters the main trough at some distance above the floor of the latter, although the surface of the two ice streams may be accordant. Glacier troughs can be studied best where large glaciers have recently melted completely away, as is the case in many valleys of the mountains of the western United States and of central and northern Europe (Fig. 114). The typical glacier trough, as shown in such examples, is U-shaped, with a broad, flat floor, and high, steep walls. Its walls are little broken by projecting spurs and lateral ravines. It is as if a V- valley cut by a river had afterwards been gouged deeper with a gigantic chisel, widening the floor to the width of the chisel blade, cutting back the spurs, and smoothing and steepening the sides. A river valley could only be as wide-floored as this after it had long been worn down to grade.

The floor of a glacier trough may not be graded; it is often interrupted by irregular steps perhaps hundreds and even a thousand feet in height, over which the stream that now drains the valley tumbles in waterfalls. Reaches between the steps are often basined. Lakelets may occupy hollows excavated in solid rock, and other lakes may be held behind terminal moraines left as dams across the valley at pauses in the retreat of the glacier.

FJORDS are glacier troughs now occupied in part or wholly by the sea, either because they were excavated by a tide glacier to their present depth below sea level, or because of a submergence of the land. Their characteristic form is that of a long, deep, narrow bay with steep rock walls and basined floor. Fjords are found only in regions which have suffered glaciation, such as Norway and Alaska.

HANGING VALLEYS. These are lateral valleys which open on their main valley some distance above its floor. They are conspicuous features of glacier troughs from which the ice has vanished; for the trunk glacier in widening and deepening its channel cut its bed below the bottoms of the lateral valleys.

Since the mouths of hanging valleys are suspended on the walls of the glacier trough, their streams are compelled to plunge down its steep, high sides in waterfalls. Some of the loftiest and most beautiful waterfalls of the world leap from hanging valleys,-- among them the celebrated Staubbach of the Lauterbrunnen valley of Switzerland, and those of the fjords of Norway and Alaska.

Hanging valleys are found also in river gorges where the smaller tributaries have not been able to keep pace with a strong master stream in cutting down their beds. In this case, however, they are a mark of extreme youth; for, as the trunk stream approaches grade and its velocity and power to erode its bed decrease, the side streams soon cut back their falls and wear their beds at their mouths to a common level with that of the main river. The Grand Canyon of the Colorado must be reckoned a young valley. At its base it narrows to scarcely more than the width of the river, and yet its tributaries, except the very smallest, enter it at a common level.

Why could not a wide-floored valley, such as a glacier trough, with hanging valleys opening upon it, be produced in the normal development of a river valley?

THE TROUGHS OF YOUNG AND OF MATURE GLACIERS. The features of a glacier trough depend much on the length of time the preexisting valley was occupied with ice. During the infancy of a glacier, we may believe, the spurs of the valley which it fills are but little blunted and its bed is but little broken by steps. In youth the glacier develops icefalls, as a river in youth develops waterfalls, and its bed becomes terraced with great stairs. The mature glacier, like the mature river, has effaced its falls and smoothed its bed to grade. It has also worn back the projecting spurs of its valley, making itself a wide channel with smooth sides. The bed of a mature glacier may form a long basin, since it abrades most in its upper and middle course, where its weight and motion are the greatest. Near the terminus, where weight and motion are the least, it erodes least, and may instead deposit a sheet of ground moraine, much as a river builds a flood plain in the same part of its course as it approaches maturity. The bed of a mature glacier thus tends to take the form of a long, relatively narrow basin, across whose lower end may be stretched the dam of the terminal moraine. On the disappearance of the ice the basin is rilled with a long, narrow lake, such as Lake Chelan in Washington and many of the lakes in the Highlands of Scotland.

Piedmont glaciers apparently erode but little. Beneath their lake-like expanse of sluggish or stagnant ice a broad sheet of ground moraine is probably being deposited.

Cirques and glaciated valleys rapidly lose their characteristic forms after the ice has withdrawn. The weather destroys all smoothed, polished, and scored surfaces which are not protected beneath glacial deposits. The oversteepened sides of the trough are graded by landslips, by talus slopes, and by alluvial cones. Morainic heaps of drift are dissected and carried away. Hanging valleys and the irregular bed of the trough are both worn down to grade by the streams which now occupy them. The length of time since the retreat of the ice from a mountain valley may thus be estimated by the degree to which the destruction of the characteristic features of the glacier trough has been carried.

In Figure 104 what characteristics of a glacier trough do you notice? What inference do you draw as to the former thickness of the glacier?

Name all the evidences you would expect to find to prove the fact that in the recent geological past the valleys of the Alps contained far larger glaciers than at present, and that on the north of the Alps the ice streams united in a piedmont glacier which extended across the plains of Switzerland to the sides of the Jura Mountains.

THE RELATIVE IMPORTANCE OF GLACIERS AND OF RIVERS. Powerful as glaciers are, and marked as are the land forms which they produce, it is easy to exaggerate their geological importance as compared with rivers. Under present climatic conditions they are confined to lofty mountains or polar lands. Polar ice sheets are permanent only so long as the lands remain on which they rest. Mountain glaciers can stay only the brief time during which the ranges continue young and high. As lofty mountains, such as the Selkirks and the Alps, are lowered by frost and glacier ice, the snowfall will decrease, the line of permanent snow will rise, and as the mountain hollows in which snow may gather are worn beneath the snow line, the glaciers must disappear.

Under present climatic conditions the work of glaciers is therefore both local and of short duration.

Even the glacial epoch, during which vast ice sheets deposited drift over northeastern North America, must have been brief as well as recent, for many lofty mountains, such as the Rockies and the Alps, still bear the marks of great glaciers which then filled their valleys. Had the glacial epoch been long, as the earth counts time, these mountains would have been worn low by ice; had the epoch been remote, the marks of glaciation would already have been largely destroyed by other agencies.

On the other hand, rivers are well-nigh universally at work over the land surfaces of the globe, and ever since the dry land appeared they have been constantly engaged in leveling the continents and in delivering to the seas the waste which there is built into the stratified rocks.

ICEBERGS. Tide glaciers, such as those of Greenland and Alaska, are able to excavate their beds to a considerable distance below sea level. From their fronts the buoyancy of sea water raises and breaks away great masses of ice which float out to sea as icebergs. Only about one seventh of a mass of glacier ice floats above the surface, and a berg three hundred feet high may be estimated to have been detached from a glacier not less than two thousand feet thick where it met the sea.

Icebergs transport on their long journeys whatever drift they may have carried when part of the glacier, and scatter it, as they melt, over the ocean floor. In this way pebbles torn by the inland ice from the rocks of the interior of Greenland and glaciated during their carriage in the ground moraine are dropped at last among the oozes of the bottom of the North Atlantic.

CHAPTER VI

THE WORK OF THE WIND

We are now to study the geological work of the currents of the atmosphere, and to learn how they erode, and transport and deposit waste as they sweep over the land. Illustrations of the wind's work are at hand in dry weather on any windy day.

Clouds of dust are raised from the street and driven along by the gale. Here the roadway is swept bare; and there, in sheltered places, the dust settles in little windrows. The erosive power of waste-laden currents of air is suggested as the sharp grains of flying sand sting one's face or clatter against the window. In the country one sometimes sees the dust whirled in clouds from dry, plowed fields in spring and left in the lee of fences in small drifts resembling in form those of snow in winter.

THE ESSENTIAL CONDITIONS for the wind's conspicuous work are illustrated in these simple examples; they are aridity and the absence of vegetation. In humid climates these conditions are only rarely and locally met; for the most part a thick growth of vegetation protects the moist soil from the wind with a cover of leaves and stems and a mattress of interlacing roots. But in arid regions either vegetation is wholly lacking, or scant growths are found huddled in detached clumps, leaving interspaces of unprotected ground (Fig. 119). Here, too, the mantle of waste, which is formed chiefly under the action of temperature changes, remains dry and loose for long periods. Little or no moisture is present to cause its particles to cohere, and they are therefore readily lifted and drifted by the wind.

TRANSPORTATION BY THE WIND

In the desert the finer waste is continually swept to and fro by the ever-shifting wind. Even in quiet weather the air heated by contact with the hot sands rises in whirls, and the dust is lifted in stately columns, sometimes as much as one thousand feet in height, which march slowly across the plain. In storms the sand is driven along the ground in a continuous sheet, while the air is tilled with dust. Explorers tell of sand storms in the deserts of central Asia and Africa, in which the air grows murky and suffocating. Even at midday it may become dark as night, and nothing can be heard except the roar of the blast and the whir of myriads of grains of sand as they fly past the ear.

Sand storms are by no means uncommon in the arid regions of the western United States. In a recent year, six were reported from Yuma, Arizona. Trains on transcontinental railways are occasionally blockaded by drifting sand, and the dust sifts into closed passenger coaches, covering the seats and floors. After such a storm thirteen car loads of sand were removed from the platform of a station on a western railway.

DUST FALLS. Dust launched by upward-whirling winds on the swift currents of the upper air is often blown for hundreds of miles beyond the arid region from which it was taken. Dust falls from western storms are not unknown even as far east as the Great Lakes. In 1896 a "black snow" fell in Chicago, and in another dust storm in the same decade the amount of dust carried in the air over Rock Island, Ill., was estimated at more than one thousand tons to the cubic mile.

In March, 1901, a cyclonic storm carried vast quantities of dust from the Sahara northward across the Mediterranean to fall over southern and central Europe. On March 8th dust storms raged in southern Algeria; two days later the dust fell in Italy; and on the 11th it had reached central Germany and Denmark. It is estimated that in these few days one million eight hundred thousand tons of waste were carried from northern Africa and deposited on European soil.

We may see from these examples the importance of the wind as an agent of transportation, and how vast in the aggregate are the loads which it carries. There are striking differences between air and water as carriers of waste. Rivers flow in fixed and narrow channels to definite goals. The channelless streams of the air sweep across broad areas, and, shifting about continually, carry their loads back and forth, now in one direction and now in another.

WIND DEPOSITS

The mantle of waste of deserts is rapidly sorted by the wind. The coarser rubbish, too heavy to be lifted into the air, is left to strew wide tracts with residual gravels (Fig. 120). The sand derived from the disintegration of desert rocks gathers in vast fields. About one eighth of the surface of the Sahara is said to be thus covered with drifting sand. In desert mountains, as those of Sinai, it lies like fields of snow in the high valleys below the sharp peaks. On more level tracts it accumulates in seas of sand, sometimes, as in the deserts of Arabia, two hundred and more feet deep.

DUNES. The sand thus accumulated by the wind is heaped in wavelike hills called dunes. In the desert of northwestern India, where the prevalent wind is of great strength, the sand is laid in longitudinal dunes, i.e. in stripes running parallel with the direction of the wind; but commonly dunes lie, like ripple marks, transverse to the wind current. On the windward side they show a long, gentle slope, up which grains of sand can readily be moved; while to the lee their slope is frequently as great as the angle of repose (Fig. 122). Dunes whose sands are not fixed by vegetation travel slowly with the wind; for their material is ever shifted forward as the grains are driven up the windward slope and, falling over the crest, are deposited in slanting layers in the quiet of the lee.

Like river deposits, wind-blown sands are stratified, since they are laid by currents of air varying in intensity, and therefore in transporting power, which carry now finer and now coarser materials and lay them down where their velocity is checked (Fig. 123). Since the wind varies in direction, the strata dip in various directions. They also dip at various angles, according to the inclination of the surface on which they were laid.

Dunes occur not only in arid regions, but also wherever loose sand lies unprotected by vegetation from the wind. From the beaches of sea and lake shores the wind drives inland the surface sand left dry between tides and after storms, piling it in dunes which may invade forests and fields and bury villages beneath their slowly advancing waves. On flood plains during summer droughts river deposits are often worked over by the wind; the sand is heaped in hummocks and much of the fine silt is caught and held by the forests and grassy fields of the bordering hills.

The sand of shore dunes differs little in composition and the shape of its grains from that of the beach from which it was derived. But in deserts, by the long wear of grain on grain as they are blown hither and thither by the wind, all soft minerals are ground to powder and the sand comes to consist almost wholly of smooth round grams of hard quartz.

Some marine sandstones, such as the St. Peter sandstone of the upper Mississippi valley, are composed so entirely of polished spherules of quartz that it has been believed by some that their grains were long blown about in ancient deserts before they were deposited in the sea.

DUST DEPOSITS. As desert sands are composed almost wholly of quartz, we may ask what has become of the softer minerals of which the rocks whose disintegration has supplied the sand were in part, and often in large part, composed. The softer minerals have been ground to powder, and little by little the quartz sand also is worn by attrition to fine dust. Yet dust deposits are scant and few in great deserts such as the Sahara. The finer waste is blown beyond its limits and laid in adjacent oceans, where it adds to the muds and oozes of their floors, and on bordering steppes and forest lands, where it is bound fast by vegetation and slowly accumulates in deposits of unstratified loose yellow earth. The fine waste of the Sahara has been identified in dredgings from the bottom of the Atlantic Ocean, taken hundreds of miles from the coast of Africa.

LOESS. In northern China an area as large as France is deeply covered with a yellow pulverulent earth called loess (German, loose), which many consider a dust deposit blown from the great Mongolian desert lying to the west. Loess mantles the recently uplifted mountains to the height of eight thousand feet and descends on the plains nearly to sea level. Its texture and lack of stratification give it a vertical cleavage; hence it stands in steep cliffs on the sides of the deep and narrow trenches which have been cut in it by streams.

On loess hillsides in China are thousands of villages whose eavelike dwellings have been excavated in this soft, yet firm, dry loam. While dust falls are common at the present time in this region, the loess is now being rapidly denuded by streams, and its yellow silt gives name to the muddy Hwang-ho (Yellow River), and to the Yellow Sea, whose waters it discolors for scores of miles from shore.

Wind deposits both of dust and of sand may be expected to contain the remains of land shells, bits of wood, and bones of land animals, testifying to the fact that they were accumulated in open air and not in the sea or in bodies of fresh water.

WIND EROSION

Sand-laden currents of air abrade and smooth and polish exposed rock surfaces, acting in much the same way as does the jet of steam fed with sharp sand, which is used in the manufacture of ground glass. Indeed, in a single storm at Cape Cod a plate glass of a lighthouse was so ground by flying sand that its transparency was destroyed and its removal made necessary.

Telegraph poles and wires whetted by wind-blown sands are destroyed within a few years. In rocks of unequal resistance the harder parts are left in relief, while the softer are etched away. Thus in the pass of San Bernardino, Cal., through which strong winds stream from the west, crystals of garnet are left projecting on delicate rock fingers from the softer rock in which they were imbedded.

Wind-carved pebbles are characteristically planed, the facets meeting along a summit ridge or at a point like that of a pyramid. We may suppose that these facets were ground by prevalent winds from certain directions, or that from time to time the stone was undermined and rolled over as the sand beneath it was blown away on the windward side, thus exposing fresh surfaces to the driving sand. Such wind-carved pebbles are sometimes found in ancient rocks and may be accepted as evidence that the sands of which the rocks are composed were blown about by the wind.

DEFLATION. In the denudation of an arid region, wind erosion is comparatively ineffective as compared with deflation (Latin, de, from; flare, to blow),--a term by which is meant the constant removal of waste by the wind, leaving the rocks bare to the continuous attack of the weather. In moist climates denudation is continually impeded by the mantle of waste and its cover of vegetation, and the land surface can be lowered no faster than the waste is removed by running water. Deep residual soils come to protect all regions of moderate slope, concealing from view the rock structure, and the various forms of the land are due more to the agencies of erosion and transportation than to differences in the resistance of the underlying rocks.

But in arid regions the mantle is rapidly removed, even from well-nigh level plains and plateaus, by the sweep of the wind and the wash of occasional rains. The geological structure of these regions of naked rock can be read as far as the eye can see, and it is to this structure that the forms of the land are there largely due. In a land mass of horizontal strata, for example, any softer surface rocks wear down to some underlying, resistant stratum, and this for a while forms the surface of a level plateau (Fig. 129). The edges of the capping layer, together with those of any softer layers beneath it, wear back in steep cliffs, dissected by the valleys of wet-weather streams and often swept bare to the base by the wind. As they are little protected by talus, which commonly is removed about as fast as formed, these escarpments and the walls of the valleys retreat indefinitely, exposing some hard stratum beneath which forms the floor of a widening terrace.

The high plateaus of northern Arizona and southern Utah, north of the Grand Canyon of the Colorado River, are composed of stratified rocks more than ten thousand feet thick and of very gentle inclination northward. From the broad plat form in which the canyon has been cut rises a series of gigantic stairs, which are often more than one thousand feet high and a score or more of miles in breadth. The retreating escarpments, the cliffs of the mesas and buttes which they have left behind as outliers, and the walls of the ravines are carved into noble architectural forms-- into cathedrals, pyramids, amphitheaters, towers, arches, and colonnades--by the processes of weathering aided by deflation. It is thus by the help of the action of the wind that great plateaus in arid regions are dissected and at last are smoothed away to waterless plains, either composed of naked rock, or strewed with residual gravels, or covered with drifting residual sand.

The specific gravity of air is 1/823 that of water. How does this fact affect the weight of the material which each can carry at the same velocity?

If the rainfall should lessen in your own state to from five to ten inches a year, what changes would take place in the vegetation of the country? in the soil? in the streams? in the erosion of valleys? in the agencies chiefly at work in denuding the land?

In what way can a wind-carved pebble be distinguished from a river-worn pebble? from a glaciated pebble?

CHAPTER VII

THE SEA AND ITS SHORES

We have already seen that the ocean is the goal at which the waste of the land arrives. The mantle of rock waste, creeping down slopes, is washed to the sea by streams, together with the material which the streams have worn from their beds and that dissolved by underground waters. In arid regions the winds sweep waste either into bordering oceans or into more humid regions where rivers take it up and carry it on to the sea. Glaciers deliver the load of their moraines either directly to the sea or leave it for streams to transport to the same goal. All deposits made on the land, such as the flood plains of rivers, the silts of lake beds, dune sands, and sheets of glacial drift, mark but pauses in the process which is to bring all the materials of the land now above sea level to rest upon the ocean bed.

But the sea is also at work along all its shores as an agent of destruction, and we must first take up its work in erosion before we consider how it transports and deposits the waste of the land.

SEA EROSION

THE SEA CLIFF AND THE ROCK BENCH. On many coasts the land fronts the ocean in a line of cliffs. To the edge of the cliffs there lead down valleys and ridges, carved by running water, which, if extended, would meet the water surface some way out from shore. Evidently they are now abruptly cut short at the present shore line because the land has been cut back.

Along the foot of the cliff lies a gently shelving bench of rock, more or less thickly veneered with sand and shingle. At low tide its inner margin is laid bare, but at high tide it is covered wholly, and the sea washes the base of the cliffs. A notch, of which the SEA CLIFF and the ROCK BENCH are the two sides, has been cut along the shore.

WAVES. The position of the rock bench, with its inner margin slightly above low tide, shows that it has been cut by some agent which acts like a horizontal saw set at about sea level. This agent is clearly the surface agitation of the water; it is the wind-raised wave.

As a wave comes up the shelving bench the crest topples forward and the wave "breaks," striking a blow whose force is measured by the momentum of all its tons of falling water. On the coast of Scotland the force of the blows struck by the waves of the heaviest storms has sometimes exceeded three tons to the square foot. But even a calm sea constantly chafes the shore. It heaves in gentle undulations known as the ground swell, the result of storms perhaps a thousand miles distant, and breaks on the shore in surf.

The blows of the waves are not struck with clear water only, else they would have little effect on cliffs of solid rock. Storm waves arm themselves with the sand and gravel, the cobbles, and even the large bowlders which lie at the base of the cliff, and beat against it with these hammers of stone.

Where a precipice descends sheer into deep water, waves swash up and down the face of the rocks but cannot break and strike effective blows. They therefore erode but little until the talus fallen from the cliff is gradually built up beneath the sea to the level at which the waves drag bottom upon it and break.

Compare the ways in which different agents abrade. The wind lightly brushes sand and dust over exposed surfaces of rock. Running water sweeps fragments of various sizes along its channels, holding them with a loose hand. Glacial ice grinds the stones of its ground moraine against the underlying rock with the pressure of its enormous weight. The wave hurls fragments of rock against the sea cliff, bruising and battering it by the blow. It also rasps the bench as it drags sand and gravel to and fro upon it.

WEATHERING OF SEA CLIFFS. The sea cliff furnishes the weapons for its own destruction. They are broken from it not only by the wave but also by the weather. Indeed the sea cliff weathers more rapidly, as a rule, than do rock ledges inland. It is abundantly wet with spray. Along its base the ground water of the neighboring land finds its natural outlet in springs which under mine it. Moreover, it is unprotected by any shield of talus. Fragments of rock as they fall from its face are battered to pieces by the waves and swept out to sea. The cliff is thus left exposed to the attack of the weather, and its retreat would be comparatively rapid for this reason alone.

Sea cliffs seldom overhang, but commonly, as in Figure 134, slope seaward, showing that the upper portion has retreated at a more rapid rate than has the base. Which do you infer is on the whole the more destructive agent, weathering or the wave?

Draw a section of a sea cliff cut in well jointed rocks whose joints dip toward the land. Draw a diagram of a sea cliff where the joints dip toward the sea.

SEA CAVES. The wave does not merely batter the face of the cliff. Like a skillful quarryman it inserts wedges in all natural fissures, such as joints, and uses explosive forces. As a wave flaps against a crevice it compresses the air within with the sudden stroke; as it falls back the air as suddenly expands. On lighthouses heavily barred doors have been burst outward by the explosive force of the air within, as it was released from pressure when a partial vacuum was formed by the refluence of the wave. Where a crevice is filled with water the entire force of the blow of the wave is transmitted by hydraulic pressure to the sides of the fissure. Thus storm waves little by little pry and suck the rock loose, and in this way, and by the blows which they strike with the stones of the beach, they quarry out about a joint, or wherever the rock may be weak, a recess known as a SEA CAVE, provided that the rock above is coherent enough to form a roof. Otherwise an open chasm results.

BLOWHOLES AND SEA ARCHES. As a sea cave is drilled back into the rock, it may encounter a joint or crevice opened to the surface by percolating water. The shock of the waves soon enlarges this to a blowhole, which one may find on the breezy upland, perhaps a hundred yards and more back from the cliff's edge. In quiet weather the blowhole is a deep well; in storm it plays a fountain as the waves drive through the long tunnel below and spout their spray high in air in successive jets. As the roof of the cave thus breaks down in the rear, there may remain in front for a while a sea arch, similar to the natural bridges of land caverns.

STACKS AND WAVE-CUT ISLANDS. As the sea drives its tunnels and open drifts into the cliff, it breaks through behind the intervening portions and leaves them isolated as stacks, much as monuments are detached from inland escarpments by the weather; and as the sea cliff retreats, these remnant masses may be left behind as rocky islets. Thus the rock bench is often set with stacks, islets in all stages of destruction, and sunken reefs, all wrecks of the land testifying to its retreat before the incessant attack of the waves.

COVES. Where zones of soft or closely jointed rock outcrop along a shore, or where minor water courses conic down to the sea and aid in erosion, the shore is worn back in curved reentrants called coves; while the more resistant rocks on either hand are left projecting as headlands (Fig. 139). After coves are cut back a short distance by the waves, the headlands come to protect them, as with breakwaters, and prevent their indefinite retreat. The shore takes a curve of equilibrium, along which the hard rock of the exposed headland and the weak rock of the protected cove wear back at an equal rate.

RATE OF RECESSION. The rate at which a shore recedes depends on several factors. In soft or incoherent rocks exposed to violent storms the retreat is so rapid as to be easily measured. The coast of Yorkshire, England, whose cliffs are cut in glacial drift, loses seven feet a year on the average, and since the Norman conquest a strip a mile wide, with farmsteads and villages and historic seaports, has been devoured by the sea. The sandy south shore of Martha's Vineyard wears back three feet a year. But hard rocks retreat so slowly that their recession has seldom been measured by the records of history.

SHORE DRIFT

BOWLDER AND PEBBLE BEACHES. About as fast as formed the waste of the sea cliff is swept both along the shore and out to sea. The road of waste along shore is the BEACH. We may also define the beach as the exposed edge of the sheet of sediment formed by the carriage of land waste out to sea. At the foot of sea cliffs, where the waves are pounding hardest, one commonly finds the rock bench strewn on its inner margin with large stones, dislodged by the waves and by the weather and some-what worn on their corners and edges. From this BOWLDER BEACH the smaller fragments of waste from the cliff and the fragments into which the bowlders are at last broken drift on to more sheltered places and there accumulate in a PEBBLE BEACH, made of pebbles well rounded by the wear which they have suffered. Such beaches form a mill whose raw material is constantly supplied by the cliff. The breakers of storms set it in motion to a depth of several feet, grinding the pebbles together with a clatter to be heard above the roar of the surf. In such a rock crusher the life of a pebble is short. Where ships have stranded on our Atlantic coast with cargoes of hard-burned brick or of coal, a year of time and a drift of five miles along the shore have proved enough to wear brick and coal to powder. At no great distance from their source, therefore, pebble beaches give place to beaches of sand, which occupy the more sheltered reaches of the shore.

SAND BEACHES. The angular sand grains of various minerals into which pebbles are broken by the waves are ground together under the beating surf and rounded, and those of the softer minerals are crushed to powder. The process, however, is a slow one, and if we study these sand grains under a lens we may be surprised to see that, though their corners and edges have been blunted, they are yet far from the spherical form of the pebbles from which they were derived. The grains are small, and in water they have lost about half their weiglit in air; the blows which they strike one another are therefore weak. Besides, each grain of sand of the wet beach is protected by a cushion of water from the blows of its neighbors.

The shape and size of these grains and the relative proportion of grains of the softer minerals which still remain give a rough measure of the distance in space and time which they have traveled from their source. The sand of many beaches, derived from the rocks of adjacent cliffs or brought in by torrential streams from neighboring highlands, is dark with grains of a number of minerals softer than quartz. The white sand of other beaches, as those of the east coast of Florida, is almost wholly composed of quartz grains; for in its long travel down the Atlantic coast the weaker minerals have been worn to powder and the hardest alone survive.

How does the absence of cleavage in quartz affect the durability of quartz sand?

HOW SHORE DRIFT MIGRATES. It is under the action of waves and currents that shore drift migrates slowly along a coast. Where waves strike a coast obliquely they drive the waste before them little by little along the shore. Thus on a north-south coast, where the predominant storms are from the northeast, there will be a migration of shore drift southwards.

All shores are swept also by currents produced by winds and tides. These are usually far too gentle to transport of themselves the coarse materials of which beaches are made. But while the wave stirs the grains of sand and gravel, and for a moment lifts them from the bottom, the current carries them a step forward on their way. The current cannot lift and the wave cannot carry, but together the two transport the waste along the shore. The road of shore drift is therefore the zone of the breaking waves.

THE BAY-HEAD BEACH. As the waste derived from the wear of waves and that brought in by streams is trailed along a coast it assumes, under varying conditions, a number of distinct forms. When swept into the head of a sheltered bay it constitutes the bay-head beach. By the highest storm waves the beach is often built higher than the ground immediately behind it, and forms a dam inclosing a shallow pond or marsh.

THE BAY BAR. As the stream of shore drift reaches the mouth of a bay of some size it often occurs that, instead of turning in, it sets directly across toward the opposite headland. The waste is carried out from shore into the deeper waters of the bay mouth; where it is no longer supported by the breaking waves, and sinks to the bottom. The dump is gradually built to the surface as a stubby spur, pointing across the bay, and as it reaches the zone of wave action current and wave can now combine to carry shore drift along it, depositing their load continually at the point of the spur. An embankment is thus constructed in much the same manner as a railway fill, which, while it is building, serves as a roadway along which the dirt from an adjacent cut is carted to be dumped at the end. When the embankment is completed it bridges the bay with a highway along which shore drift now moves without interruption, and becomes a bay bar.

INCOMPLETE BAY BARS. Under certain conditions the sea cannot carry out its intention to bridge a bay. Rivers discharging in bays demand open way to the ocean. Strong tidal currents also are able to keep open channels scoured by their ebb and flow. In such cases the most that land waste can do is to build spits and shoals, narrowing and shoaling the channel as much as possible. Incomplete bay bars sometimes have their points recurved by currents setting at right angles to the stream of shore drift and are then classified as HOOKS (Fig. 142).

SAND REEFS. On low coasts where shallow water extends some distance out, the highway of shore drift lies along a low, narrow ridge, termed the sand reef, separated from the land by a narrow stretch of shallow water called the LAGOON. At intervals the reef is held open by INLETS,--gaps through which the tide flows and ebbs, and by which the water of streams finds way to the sea.

No finer example of this kind of shore line is to be found in the world than the coast of Texas. From near the mouth of the Rio Grande a continuous sand reef draws its even curve for a hundred miles to Corpus Christi Pass, and the reefs are but seldom interrupted by inlets as far north as Galveston Harbor. On this coast the tides are variable and exceptionally weak, being less than one foot in height, while the amount of waste swept along the shore is large. The lagoon is extremely shallow, and much of it is a mud flat too shoal for even small boats. On the coast of New Jersey strong tides are able to keep open inlets at intervals of from two to twenty miles in spite of a heavy alongshore drift.

Sand reefs are formed where the water is so shallow near shore that storm waves cannot run in it and therefore break some distance out from land. Where storm waves first drag bottom they erode and deepen the sea floor, and sweep in sediment as far as the line where they break. Here, where they lose their force, they drop their load and beat up the ridge which is known as the sand reef when it reaches the surface.

SHORES OF ELEVATION AND DEPRESSION

Our studies have already brought to our notice two distinct forms of strand lines,--one the high, rocky coast cut back to cliffs by the attack of the waves, and the other the low, sandy coast where the waves break usually upon the sand reef. To understand the origin of these two types we must know that the meeting place of sea and land is determined primarily by movements of the earth's crust. Where a coast land emerges the--shore line moves seaward; where it is being submerged the shore line advances on the land.

SHORES OF ELEVATION. The retreat of the sea, either because of a local uplift of the land or for any other reason, such as the lowering of any portion of ocean bottom, lays bare the inner margin of the sea floor. Where the sea floor has long received the waste of the land it has been built up to a smooth, subaqueous plain, gently shelving from the land. Since the new shore line is drawn across this even surface it is simple and regular, and is bordered on the one side by shallow water gradually deepening seaward, and on the other by low land composed of material which has not yet thoroughly consolidated to firm rock. A sand reef is soon beaten up by the waves, and for some time conditions will favor its growth. The loss of sand driven into the lagoon beyond, and of that ground to powder by the surf and carried out to sea, is more than made up by the stream of alongshore drift, and especially by the drag of sediments to the reef by the waves as they deepen the sea floor on its seaward side.

Meanwhile the lagoon gradually fills with waste from the reef and from the land. It is invaded by various grasses and reeds which have learned to grow in salt and brackish water; the marsh, laid bare only at low tide, is built above high tide by wind drift and vegetable deposits, and becomes a meadow, soldering the sand reef to the mainland.

While the lagoon has been filling, the waves have been so deepening the sea floor off the sand reef that at last they are able to attack it vigorously. They now wear it back, and, driving the shore line across the lagoon or meadow, cut a line of low cliffs on the mainland. Such a shore is that of Gascony in southwestern France,--a low, straight, sandy shore, bordered by dunes and unprotected by reefs from the attack of the waves of the Bay of Biscay.

We may say, then, that on shores of elevation the presence of sand reefs and lagoons indicates the stage of youth, while the absence of these features and the vigorous and unimpeded attack by the sea upon the mainland indicate the stage of maturity. Where much waste is brought in by rivers the maturity of such a coast may be long delayed. The waste from the land keeps the sea shallow offshore and constantly renews the sand reef. The energy of the waves is consumed in handling shore drift, and no energy is left for an effective attack upon the land. Indeed, with an excessive amount of waste brought down by streams the land may be built out and encroach temporarily upon the sea; and not until long denudation has lowered the land, and thus decreased the amount of waste from it, may the waves be able to cut through the sand reef and thus the coast reach maturity.

SHORES OF DEPRESSION

Where a coastal region is undergoing submergence the shore line moves landward. The horizontal plane of the sea now intersects an old land surface roughened by subaerial denudation. The shore line is irregular and indented in proportion to the relief of the land and the amount of the submergence which the land has suffered. It follows up partially submerged valleys, forming bays, and bends round the divides, leaving them to project as promontories and peninsulas. The outlines of shores of depression are as varied as are the forms of the land partially submerged. We give a few typical illustrations.

The characteristics of the coast of Maine are due chiefly to the fact that a mountainous region of hard rocks, once worn to a peneplain, and after a subsequent elevation deeply dissected by north-south valleys, has subsided, the depression amounting on its southern margin to as much as six hundred feet below sea level. Drowned valleys penetrate the land in long, narrow bays, and rugged divides project in long, narrow land arms prolonged seaward by islands representing the high portions of their extremities. Of this exceedingly ragged shore there are said to be two thousand miles from the New Brunswick boundary as far west as Portland,--a straight-line distance of but two hundred miles. Since the time of its greatest depression the land is known to have risen some three hundred feet; for the bays have been shortened, and the waste with which their floors were strewn is now in part laid bare as clay plains about the bay heads and in narrow selvages about the peninsulas and islands.

The coast of Dalmatia, on the Adriatic Sea, is characterized by long land arms and chains of long and narrow islands, all parallel to the trend of the coast. A region of parallel mountain ranges has been depressed, and the longitudinal valleys which lie between them are occupied by arms of the sea.

Chesapeake Bay is a branching bay due to the depression of an ancient coastal plain which, after having emerged from the sea, was channeled with broad, shallow valleys. The sea has invaded the valley of the trunk stream and those of its tributaries, forming a shallow bay whose many branches are all directed toward its axis (Fig. 146).

Hudson Bay, and the North, the Baltic, and the Yellow seas are examples where the sinking of the land has brought the sea in over low plains of large extent, thus deeply indenting the continental out-line. The rise of a few hundred feet would restore these submerged plains to the land.

THE CYCLE OF SHORES OF DEPRESSION. In its infantile stage the outline of a shore of depression depends almost wholly on the previous relief of the land, and but little on erosion by the sea. Sea cliffs and narrow benches appear where headlands and outlying islands have been nipped by the waves. As yet, little shore waste has been formed. The coast of Maine is an example of this stage.

In early youth all promontories have been strongly cliffed, and under a vigorous attack of the sea the shore of open bays may be cut back also. Sea stacks and rocky islets, caves and coves, make the shore minutely ragged. The irregularity of the coast, due to depression, is for a while increased by differential wave wear on harder and softer rocks. The rock bench is still narrow. Shore waste, though being produced in large amounts, is for the most part swept into deeper water and buried out of sight. Examples of this stage are the east coast of Scotland and the California coast near San Francisco.

Later youth is characterized by a large accumulation of shore waste. The rock bench has been cut back so that it now furnishes a good roadway for shore drift. The stream of alongshore drift grows larger and larger, filling the heads of the smaller bays with beaches, building spits and hooks, and tying islands with sand bars to the mainland. It bridges the larger bays with bay bars, while their length is being reduced as their inclosing promontories are cut back by the waves. Thus there comes to be a straight, continuous, and easy road, no longer interrupted by headlands and bays, for the transportation of waste alongshore. The Baltic coast of Germany is in this stage.

All this while streams have been busy filling with delta deposits the bays into which they empty. By these steps a coast gradually advances to MATURITY, the stage when the irregularities due to depression have been effaced, when outlying islands formed by subsidence have been planed away, and when the shore line has been driven back behind the former bay heads. The sea now attacks the land most effectively along a continuous and fairly straight line of cliffs. Although the first effect of wave wear was to increase the irregularities of the shore, it sooner or later rectifies it, making it simple and smooth. Northwestern France may be cited as an upland plain, dissected and depressed, whose coast has reached maturity.

In the OLD AGE of coasts the rock bench is cut back so far that the waves can no longer exert their full effect upon the shore. Their energy is dissipated in moving shore drift hither and thither and in abrading the bench when they drag bottom upon it. Little by little the bench is deepened by tidal currents and the drag of waves; but this process is so slow that meanwhile the sea cliffs melt down under the weather, and the bench becomes a broad shoal where waves and tides gradually work over the waste from the land to greater fineness and sweep it out to sea.

PLAINS OF MARINE ABRASION. While subaerial denudation reduces the land to baselevel, the sea is sawing its edges to WAVE BASE, i.e. the lowest limit of the wave's effective wear. The widened rock bench forms when uplifted a plain of marine abrasion, which like the peneplain bevels across strata regardless of their various inclinations and various degrees of hardness.

How may a plain of marine abrasion be expected to differ from a peneplain in its mantle of waste?

Compared with subaerial denudation, marine abrasion is a comparatively feeble agent. At the rate of five feet per century-- a higher rate than obtains on the youthful rocky, coast of Britain--it would require more than ten million years to pare a strip one hundred miles wide from the margin of a continent, a time sufficient, at the rate at which the Mississippi valley is now being worn away, for subaerial denudation to lower the lands of the globe to the level of the sea.

Slow submergence favors the cutting of a wide rock bench. The water continually deepens upon the bench; storm waves can therefore always ride in to the base of the cliffs and attack them with full force; shore waste cannot impede the onset of the waves, for it is continually washed out in deeper water below wave base.

BASAL CONGOLMERATES. As the sea marches across the land during a slow submergence, the platform is covered with sheets of sea-laid sediments. Lowest of these is a conglomerate,--the bowlder and pebble beach, widened indefinitely by the retreat of the cliffs at whose base it was formed, and preserved by the finer deposits laid upon it in the constantly deepening water as the land subsides. Such basal conglomerates are not uncommon among the ancient rocks of the land, and we may know them by their rounded pebbles and larger stones, composed of the same kind of rock as that of the abraded and evened surface on which they lie.

CHAPTER VIII

OFFSHORE AND DEEP-SEA DEPOSITS

The alongshore deposits which we have now studied are the exposed edge of a vast subaqueous sheet of waste which borders the continents and extends often for as much as two or three hundred miles from land. Soundings show that offshore deposits are laid in belts parallel to the coast, the coarsest materials lying nearest to the land and the finest farthest out. The pebbles and gravel and the clean, coarse sand of beaches give place to broad stretches of sand, which grows finer and finer until it is succeeded by sheets of mud. Clearly there is an offshore movement of waste by which it is sorted, the coarser being sooner dropped and the finer being carried farther out.

OFFSHORE DEPOSITS

The debris torn by waves from rocky shores is far less in amount than the waste of the land brought down to the sea by rivers, being only one thirty-third as great, according to a conservative estimate. Both mingle alongshore in all the forms of beach and bar that have been described, and both are together slowly carried out to sea. On the shelving ocean floor waste is agitated by various movements of the unquiet water,--by the undertow (an outward-running bottom current near the shore), by the ebb and flow of tides, by ocean currents where they approach the land, and by waves and ground swells, whose effects are sometimes felt to a depth of six hundred feet. By all these means the waste is slowly washed to and fro, and as it is thus ground finer and finer and its soluble parts are more and more dissolved, it drifts farther and farther out from land. It is by no steady and rapid movement that waste is swept from the shore to its final resting place. Day after day and century after century the grains of sand and particles of mud are shifted to and fro, winnowed and spread in layers, which are destroyed and rebuilt again and again before they are buried safe from further disturbance.

These processes which are hidden from the eye are among the most important of those with which our science has to do; for it is they which have given shape to by far the largest part of the stratified rocks of which the land is made.

THE CONTINENTAL DELTA. This fitting term has been recently suggested for the sheet of waste slowly accumulating along the borders of the continents. Within a narrow belt, which rarely exceeds two or three hundred miles, except near the mouths of muddy rivers such as the Amazon and Congo, nearly all the waste of the continent, whether worn from its surface by the weather, by streams, by glaciers, or by the wind, or from its edge by the chafing of the waves, comes at last to its final resting place. The agencies which spread the material of the continental delta grow more and more feeble as they pass into deeper and more quiet water away from shore. Coarse materials are therefore soon dropped along narrow belts near land. Gravels and coarse sands lie in thick, wedge-shaped masses which thin out seaward rapidly and give place to sheets of finer sand.

SEA MUDS. Outermost of the sediments derived from the waste of the continents is a wide belt of mud; for fine clays settle so slowly, even in sea water,--whose saltness causes them to sink much faster than they would in fresh water,--that they are wafted far before they reach a bottom where they may remain undisturbed. Muds are also found near shore, carpeting the floors of estuaries, and among stretches of sandy deposits in hollows where the more quiet water has permitted the finer silt to rest.

Sea muds are commonly bluish and consolidate to bluish shales; the red coloring matter brought from land waste--iron oxide--is altered to other iron compounds by decomposing organic matter in the presence of sea water. Yellow and red muds occur where the amount of iron oxide in the silt brought down to the sea by rivers is too great to be reduced, or decomposed, by the organic matter present.

Green muds and green sand owe their color to certain chemical changes which take place where waste from the land accumulates on the sea floor with extreme slowness. A greenish mineral called GLAUCONITE--a silicate of iron and alumina--is then formed. Such deposits, known as GREEN SAND, are now in process of making in several patches off the Atlantic coast, and are found on the coastal plain of New Jersey among the offshore deposits of earlier geological ages.

ORGANIC DEPOSITS. Living creatures swarm along the shore and on the shallows out from land as nowhere else in the ocean. Seaweed often mantles the rock of the sea cliff between the levels of high and low tide, protecting it to some degree from the blows of waves. On the rock bench each little pool left by the ebbing tide is an aquarium abounding in the lowly forms of marine life. Below low-tide level occur beds of molluscous shells, such as the oyster, with countless numbers of other humble organisms. Their harder parts--the shells of mollusks, the white framework of corals, the carapaces of crabs and other crustaceans, the shells of sea urchins, the bones and teeth of fishes--are gradually buried within the accumulating sheets of sediment, either whole or, far more often, broken into fragments by the waves.

By means of these organic remains each layer of beach deposits and those of the continental delta may contain a record of the life of the time when it was laid. Such a record has been made ever since living creatures with hard parts appeared upon the globe. We shall find it sealed away in the stratified rocks of the continents,-- parts of ancient sea deposits now raised to form the dry land. Thus we have in the traces of living creatures found in the rocks, i.e. in fossils, a history of the progress of life upon the planet.

MOLLUSCOUS SHELL DEPOSITS. The forms of marine life of importance in rock making thrive best in clear water, where little sediment is being laid, and where at the same time the depth is not so great as to deprive them of needed light, heat, and of sufficient oxygen absorbed by sea water from the air. In such clear and comparatively shallow water there often grow countless myriads of animals, such as mollusks and corals, whose shells and skeletons of carbonate of lime gradually accumulate in beds of limestone.

A shell limestone made of broken fragments cemented together is sometimes called COQUINA, a local term applied to such beds recently uplifted from the sea along the coast of Florida (Fig. 149).

OOLITIC limestone (oon, an egg; lithos, a stone) is so named from the likeness of the tiny spherules which compose it to the roe of fish. Corals and shells have been pounded by the waves to calcareous sand, and each grain has been covered with successive concentric coatings of lime carbonate deposited about it from solution.

The impalpable powder to which calcareous sand is ground by the waves settles at some distance from shore in deeper and quieter water as a limy silt, and hardens into a dense, fine-grained limestone in which perhaps no trace of fossil is found to suggest the fact that it is of organic origin.

From Florida Keys there extends south to the trough of Florida Straits a limestone bank covered by from five hundred and forty to eighteen hundred feet of water. The rocky bottom consists of limestone now slowly building from the accumulation of the remains of mollusks, small corals, sea urchins, worms with calcareous tubes, and lime-secreting seaweed, which live upon its surface.

Where sponges and other silica-secreting organisms abound on limestone banks, silica forms part of the accumulated deposit, either in its original condition, as, for example, the spicules of sponges, or gathered into concretions and layers of flint.

Where considerable mud is being deposited along with carbonate of lime there is in process of making a clayey limestone or a limy shale; where considerable sand, a sandy limestone or a limy sandstone.

CONSOLIDATION OF OFFSHORE DEPOSITS. We cannot doubt that all these loose sediments of the sea floor are being slowly consolidated to solid rock. They are soaked with water which carries in solution lime carbonate and other cementing substances. These cements are deposited between the fragments of shells and corals, the grains of sand and the particles of mud, binding them together into firm rock. Where sediments have accumulated to great thickness the lower portions tend also to consolidate under the weight of the overlying beds. Except in the case of limestones, recent sea deposits uplifted to form land are seldom so well cemented as are the older strata, which have long been acted upon by underground waters deep below the surface within the zone of cementation, and have been exposed to view by great erosion.

RIPPLE MARKS, SUN CRACKS, ETC. The pulse of waves and tidal currents agitates the loose material of offshore deposits, throwing it into fine parallel ridges called ripple marks. One may see this beautiful ribbing imprinted on beach sands uncovered by the outgoing tide, and it is also produced where the water is of considerable depth. While the tide is out the surface of shore deposits may be marked by the footprints of birds and other animals, or by the raindrops of a passing shower.

The mud of flats, thus exposed to the sun and dried, cracks in a characteristic way. Such markings may be covered over with a thin layer of sediment at the next flood tide and sealed away as a lasting record of the manner and place in which the strata were laid. In Figure 150 we have an illustration of a very ancient ripple-marked sand consolidated to hard stone, uplifted and set on edge by movements of the earth's crust, and exposed to open air after long erosion.

STRATIFICATION. For the most part the sheet of sea-laid waste is hidden from our sight. Where its edge is exposed along the shore we may see the surface markings which have just been noticed. Soundings also, and the observations made in shallow waters by divers, tell something of its surface; but to learn more of its structures we must study those ancient sediments which have been lifted from the sea and dissected by subaerial agencies. From them we ascertain that sea deposits are stratified. They lie in distinct layers which often differ from one another in thickness, in size of particles, and perhaps in color. They are parted by bedding planes, each of which represents either a change in material or a pause during which deposition ceased and the material of one layer had time to settle and become somewhat consolidated before the material of the next was laid upon it. Stratification is thus due to intermittently acting forces, such as the agitation of the water during storms, the flow and ebb of the tide, and the shifting channels of tidal currents. Off the mouths of rivers, stratification is also caused by the coarser and more abundant material brought down at time of floods being laid on the finer silt which is discharged during ordinary stages.

How stratified deposits are built up is well illustrated in the flats which border estuaries, such as the Bay of Fundy. Each advance of the tide spreads a film of mud, which dries and hardens in the air during low water before another film is laid upon it by the next incoming tidal flood. In this way the flats have been covered by a clay which splits into leaves as thin as sheets of paper.

It is in fine material, such as clays and shales and limestones, that the thinnest and most uniform layers, as well as those of widest extent, occur. On the other hand, coarse materials are commonly laid in thick beds, which soon thin out seaward and give place to deposits of finer stuff. In a general way strata are laid in well-nigh horizontal sheets, for the surface on which they are laid is generally of very gentle inclination. Each stratum, however, is lenticular, or lenslike, in form, having an area where it is thickest, and thinning out thence to its edges, where it is overlapped by strata similar in shape.

CROSS BEDDING. There is an apparent exception to this rule where strata whose upper and lower surfaces may be about horizontal are made up of layers inclined at angles which may be as high as the angle of repose. In this case each stratum grew by the addition along its edge of successive layers of sediment, precisely as does a sand bar in a river, the sand being pushed continuously over the edge and coming to rest on a sloping surface. Shoals built by strong and shifting tidal currents often show successive strata in which the cross bedding is inclined in different directions.

THICKNESS OF SEA DEPOSITS. Remembering the vast amount of material denuded from the land and deposited offshore, we should expect that with the lapse of time sea deposits would have grown to an enormous thickness. It is a suggestive fact that, as a rule, the profile of the ocean bed is that of a soup plate,--a basin surrounded by a flaring rim. On the CONTINENTAL SHELF, as the rim is called, the water is seldom more than six hundred feet in depth at the outer edge, and shallows gradually towards shore. Along the eastern coast of the United States the continental shelf is from fifty to one hundred and more miles in width; on the Pacific coast it is much narrower. So far as it is due to upbuilding, a wide continental shelf, such as that of the Atlantic coast, implies a massive continental delta thousands of feet in thickness. The coastal plain of the Atlantic states may be regarded as the emerged inner margin of this shelf, and borings made along the coast probe it to the depth of as much as three thousand feet without finding the bottom of ancient offshore deposits. Continental shelves may also be due in part to a submergence of the outer margin of a continental plateau and to marine abrasion.

DEPOSITION OF SEDIMENTS AND SUBSIDENCE. The stratified rocks of the land show in many places ancient sediments which reach a thickness which is measured in miles, and which are yet the product of well-nigh continuous deposition. Such strata may prove by their fossils and by their composition and structure that they were all laid offshore in shallow water. We must infer that, during the vast length of time recorded by the enormous pile, the floor of the sea along the coast was slowly sinking, and that the trough was constantly being filled, foot by foot, as fast as it was depressed. Such gradual, quiet movements of the earth's crust not only modify the outline of coasts, as we have seen, but are of far greater geological importance in that they permit the making of immense deposits of stratified rock.

A slow subsidence continued during long time is recorded also in the succession of the various kinds of rock that come to be deposited in the same area. As the sea transgresses the land, i.e. encroaches upon it, any given part of the sea bottom is brought farther and farther from the shore. The basal conglomerate formed by bowlder and pebble beaches comes to be covered with sheets of sand, and these with layers of mud as the sea becomes deeper and the shore more remote; while deposits of limestone are made when at last no waste is brought to the place from the now distant land, and the water is left clear for the growth of mollusks and other lime-secreting organisms.

RATE OF DEPOSITION. As deposition in the sea corresponds to denudation on the land, we are able to make a general estimate of the rate at which the former process is going on. Leaving out of account the soluble matter removed, the Mississippi is lowering its basin at the rate of one foot in five thousand years, and we may assume this as the average rate at which the earth's land surface of fifty-seven million square miles is now being denuded by the removal of its mechanical waste. But sediments from the land are spread within a zone but two or three hundred miles in width along the margin of the continents, a line one hundred thousand miles long. As the area of deposition-- about twenty-five million square miles--is about one half the area of denudation, the average rate of deposition must be twice the average rate of denudation, i.e. about one foot in twenty-five hundred years. If some deposits are made much more rapidly than this, others are made much more slowly. If they were laid no faster than the present average rate, the strata of ancient sea deposits exposed in a quarry fifty feet deep represent a lapse of at least one hundred and twenty-five thousand years, and those of a formation five hundred feet thick required for their accumulation one million two hundred and fifty thousand years.

THE SEDIMENTARY RECORD AND THE DENUDATION CYCLE. We have seen that the successive stages in a cycle of denudation, such as that by which a land mass of lofty mountains is worn to low plains, are marked each by its own peculiar land forms, and that the forms of the earlier stages are more or less completely effaced as the cycle draws toward an end. Far more lasting records of each stage are left in the sedimentary deposits of the continental delta.

Thus, in the youth of such a land mass as we have mentioned, torrential streams flowing down the steep mountain sides deliver to the adjacent sea their heavy loads of coarse waste, and thick offshore deposits of sand and gravel (Fig. 156) record the high elevation of the bordering land. As the land is worn to lower levels, the amount and coarseness of the waste brought to the sea diminishes, until the sluggish streams carry only a fine silt which settles on the ocean floor near to land in wide sheets of mud which harden into shale. At last, in the old age of the region (Fig. 157), its low plains contribute little to the sea except the soluble elements of the rocks, and in the clear waters near the land lime-secreting organisms flourish and their remains accumulate in beds of limestone. When long-weathered lands mantled with deep, well-oxidized waste are uplifted by a gradual movement of the earth's crust, and the mantle is rapidly stripped off by the revived streams, the uprise is recorded in wide deposits of red and yellow clays and sands upon the adjacent ocean floor.

Where the waste brought in is more than the waves can easily distribute, as off the mouths of turbid rivers which drain highlands near the sea, deposits are little winnowed, and are laid in rapidly alternating, shaly sandstones and sandy shales.

Where the highlands are of igneous rock, such as granite, and mechanical disintegration is going on more rapidly than chemical decay, these conditions are recorded in the nature of the deposits laid offshore. The waste swept in by streams contains much feldspar and other minerals softer and more soluble than quartz, and where the waves have little opportunity to wear and winnow it, it comes to rest in beds of sandstone in which grains of feldspar and other soft minerals are abundant. Such feldspathic sandstones are known as ARKOSE.

On the other hand, where the waste supplied to the sea comes chiefly from wide, sandy, coastal plains, there are deposited off- shore clean sandstones of well-worn grains of quartz alone. In such coastal plains the waste of the land is stored for ages. Again and again they are abandoned and invaded by the sea as from time to time the land slowly emerges and is again submerged. Their deposits are long exposed to the weather, and sorted over by the streams, and winnowed and worked over again and again by the waves. In the course of long ages such deposits thus become thoroughly sorted, and the grains of all minerals softer than quartz are ground to mud.

DEEP-SEA OOZES AND CLAYS

GLOBIGERINA OOZE. Beyond the reach of waste from the land the bottom of the deep sea is carpeted for the most part with either chalky ooze or a fine red clay. The surface waters of the warm seas swarm with minute and lowly animals belonging to the order of the Foraminifera, which secrete shells of carbonate of lime. At death these tiny white shells fall through the sea water like snowflakes in the air, and, slowly dissolving, seem to melt quite away before they can reach depths greater than about three miles. Near shore they reach bottom, but are masked by the rapid deposit of waste derived from the land. At intermediate depths they mantle the ocean floor with a white, soft lime deposit known as Globigerina ooze, from a genus of the Foraminifera which contributes largely to its formation.

RED CLAY. Below depths of from fifteen to eighteen thousand feet the ocean bottom is sheeted with red or chocolate colored clay. It is the insoluble residue of seashells, of the debris of submarine volcanic eruptions, of volcanic dust wafted by the winds, and of pieces of pumice drifted by ocean currents far from the volcanoes from which they were hurled. The red clay builds up with such inconceivable slowness that the teeth of sharks and the hard ear bones of whales may be dredged in large numbers from the deep ocean bed, where they have lain unburied for thousands of years; and an appreciable part of the clay is also formed by the dust of meteorites consumed in the atmosphere,--a dust which falls everywhere on sea and land, but which elsewhere is wholly masked by other deposits.

The dark, cold abysses of the ocean are far less affected by change than any other portion of the surface of the lithosphere. These vast, silent plains of ooze lie far below the reach of storms. They know no succession of summer and winter, or of night and day. A mantle of deep and quiet water protects them from the agents of erosion which continually attack, furrow, and destroy the surface of the land. While the land is the area of erosion, the sea is the area of deposition. The sheets of sediment which are slowly spread there tend to efface any inequalities, and to form a smooth and featureless subaqueous plain.

With few exceptions, the stratified rocks of the land are proved by their fossils and composition to have been laid in the sea; but in the same way they are proved to be offshore, shallow-water deposits, akin to those now making on continental shelves. Deep-sea deposits are absent from the rocks of the land, and we may therefore infer that the deep sea has never held sway where the continents now are,--that the continents have ever been, as now, the elevated portions of the lithosphere, and that the deep seas of the present have ever been its most depressed portions.

THE REEF-BUILDING CORALS

In warm seas the most conspicuous of rock-making organisms are the corals known as the reef builders. Floating in a boat over a coral reef, as, for example, off the south coast of Florida or among the Bahamas, one looks down through clear water on thickets of branching coral shrubs perhaps as much as eight feet high, and hemispherical masses three or four feet thick, all abloom with countless minute flowerlike coral polyps, gorgeous in their colors of yellow, orange, green, and red. In structure each tiny polyp is little more than a fleshy sac whose mouth is surrounded with petal-like tentacles, or feelers. From the sea water the polyps secrete calcium carbonate and build it up into the stony framework which supports their colonies. Boring mollusks, worms, and sponges perforate and honeycomb this framework even while its surface is covered with myriads of living polyps. It is thus easily broken by the waves, and white fragments of coral trees strew the ground beneath. Brilliantly colored fishes live in these coral groves, and countless mollusks, sea urchins, and other forms of marine life make here their home. With the debris from all these sources the reef is constantly built up until it rises to low-tide level. Higher than this the corals cannot grow, since they are killed by a few hours' exposure to the air.

When the reef has risen to wave base, the waves abrade it on the windward side and pile to leeward coral blocks torn from their foundation, filling the interstices with finer fragments. Thus they heap up along the reef low, narrow islands (Fig. 160).

Reef building is a comparatively rapid progress. It has been estimated that off Florida a reef could be built up to the surface from a depth of fifty feet in about fifteen hundred years.

CORAL LIMESTONES. Limestones of various kinds are due to the reef builders. The reef rock is made of corals in place and broken fragments of all sizes, cemented together with calcium carbonate from solution by infiltrating waters. On the island beaches coral sand is forming oolitic limestone, and the white coral mud with which the sea is milky for miles about the reef in times of storm settles and concretes into a compact limestone of finest grain. Corals have been among the most important limestone builders of the sea ever since they made their appearance in the early geological ages.

The areas on which coral limestone is now forming are large. The Great Barrier Reef of Australia, which lies off the north-eastern coast, is twelve hundred and fifty miles long, and has a width of from ten to ninety miles. Most of the islands of the tropics are either skirted with coral reefs or are themselves of coral formation.

CONDITIONS OF CORAL GROWTH. Reef-building corals cannot live except in clear salt water less, as a rule, than one hundred and fifty feet in depth, with a winter temperature not lower than 68 degrees F. An important condition also is an abundant food supply, and this is best secured in the path of the warm oceanic currents.

Coral reefs may be grouped in three classes,--fringing reefs, barrier reefs, and atolls.

FRINGING REEFS. These take their name from the fact that they are attached as narrow fringes to the shore. An example is the reef which forms a selvage about a mile wide along the northeastern coast of Cuba. The outer margin, indicated by the line of white surf, where the corals are in vigorous growth, rises from about forty feet of water. Between this and the shore lies a stretch of shoal across which one can wade at low water, composed of coral sand with here and there a clump of growing coral.

BARRIER REEFS. Reefs separated from the shore by a ship channel of quiet water, often several miles in width and sometimes as much as three hundred feet in depth, are known as barrier reefs. The seaward face rises abruptly from water too deep for coral growth. Low islands are cast up by the waves upon the reef, and inlets give place for the ebb and flow of the tides. Along the west coast of the island of New Caledonia a barrier reef extends for four hundred miles, and for a length of many leagues seldom approaches within eight miles of the shore.

ATOLLS. These are ring-shaped or irregular coral islands, or island-studded reefs, inclosing a central lagoon. The narrow zone of land, like the rim of a great bowl sunken to the water's edge, rises hardly more than twenty feet at most above the sea, and is covered with a forest of trees such as the cocoanut, whose seeds can be drifted to it uninjured from long distances. The white beach of coral sand leads down to the growing reef, on whose outer margin the surf is constantly breaking. The sea face of the reef falls off abruptly, often to depths of thousands of feet, while the lagoon varies in depth from a few feet to one hundred and fifty or two hundred, and exceptionally measures as much as three hundred and fifty feet.

THEORIES OF CORAL REEFS. Fringing reefs require no explanation, since the depth of water about them is not greater than that at which coral can grow; but barrier reefs and atolls, which may rise from depths too great for coral growth demand a theory of their origin.

Darwin's theory holds that barrier reefs and atolls are formed from fringing reefs by SUBSIDENCE. The rate of sinking cannot be greater than that of the upbuilding of the reef, since otherwise the corals would be carried below their depth and drowned. The process is illustrated in Figure 161, where v represents a volcanic island in mid ocean undergoing slow depression, and ss the sea level before the sinking began, when the island was surrounded by a fringing reef. As the island slowly sinks, the reef builds up with equal pace. It rears its seaward face more steep than the island slope, and thus the intervening space between the sinking, narrowing land and the outer margin of the reef constantly widens. In this intervening space the corals are more or less smothered with silt from the outer reef and from the land, and are also deprived in large measure of the needful supply of food and oxygen by the vigorous growth of the corals on the outer rim. The outer rim thus becomes a barrier reef and the inner belt of retarded growth is deepened by subsidence to a ship channel, s's' representing sea level at this time. The final stage, where the island has been carried completely beneath the sea and overgrown by the contracting reef, whose outer ring now forms an atoll, is represented by s"s".

In very many instances, however, atolls and barrier reefs may be explained without subsidence. Thus a barrier reef may be formed by the seaward growth of a fringing reef upon the talus of its sea face. In Figure 162 f is a fringing reef whose outer wall rises from about one hundred and fifty feet, the lower limit of the reef-building species. At the foot of this submarine cliff a talus of fallen blocks t accumulates, and as it reaches the zone of coral growth becomes the foundation on which the reef is steadily extended seaward. As the reef widens, the polyps of the circumference flourish, while those of the inner belt are retarded in their growth and at last perish. The coral rock of the inner belt is now dissolved by sea water and scoured out by tidal currents until it gives place to a gradually deepening ship channel, while the outer margin is left as a barrier reef.

In much the same way atolls may be built on any shoal which lies within the zone of coral growth. Such shoals may be produced when volcanic islands are leveled by waves and ocean currents, and when submarine plateaus, ridges, and peaks are built up by various organic agencies, such as molluscous and foraminiferal shell deposits. The reef-building corals, whose eggs are drifted widely over the tropic seas by ocean currents, colonize such submarine foundations wherever the conditions are favorable for their growth. As the reef approaches the surface the corals of the inner area are smothered by silt and starved, and their Submarine Volcanic Peak hard parts are dissolved and scoured away; while those of the circumference, with abundant food supply, nourish and build the ring of the atoll. Atolls may be produced also by the backward drift of sand from either end of a crescentic coral reef or island, the spits uniting in the quiet water of the lee to inclose a lagoon. In the Maldive Archipelago all gradations between crescent-shaped islets and complete atoll rings have been observed.

In a number of instances where coral reefs have been raised by movements of the earth's crust, the reef formation is found to be a thin veneer built upon a foundation of other deposits. Thus Christmas Island, in the Indian Ocean, is a volcanic pile rising eleven hundred feet above sea level and fifteen thousand five hundred feet above the bottom of the sea. The summit is a plateau surrounded by a rim of hills of reef formation, which represent the ring of islets of an ancient atoll. Beneath the reef are thick beds of limestone, composed largely of the remains of foraminifers, which cover the lavas and fragraental materials of the old submarine volcano.

Among the ancient sediments which now form the stratified rocks of the land there occur many thin reef deposits, but none are known of the immense thickness which modern reefs are supposed to reach according to the theory of subsidence.

Barrier and fringing reefs are commonly interrupted off the mouths of rivers. Why?

SUMMARY. We have seen that the ocean bed is the goal to which the waste of the rocks of the land at last arrives. Their soluble parts, dissolved by underground waters and carried to the sea by rivers, are largely built up by living creatures into vast sheets of limestone. The less soluble portions--the waste brought in by streams and the waste of the shore--form the muds and sands of continental deltas. All of these sea deposits consolidate and harden, and the coherent rocks of the land are thus reconstructed on the ocean floor. But the destination is not a final one. The stratified rocks of the land are for the most part ancient deposits of the sea, which have been lifted above sea level; and we may believe that the sediments now being laid offshore are the "dust of continents to be," and will some time emerge to form additions to the land. We are now to study the movements of the earth's crust which restore the sediments of the sea to the light of day, and to whose beneficence we owe the habitable lands of the present.

PART II

INTERNAL GEOLOGICAL AGENCIES

CHAPTER IX

MOVEMENTS OF THE EARTH'S CRUST

The geological agencies which we have so far studied--weathering, streams, underground waters, glaciers, winds, and the ocean--all work upon the earth from without, and all are set in motion by an energy external to the earth, namely, the radiant energy of the sun. All, too, have a common tendency to reduce the inequalities of the earth's surface by leveling the lands and strewing their waste beneath the sea.

But despite the unceasing efforts of these external agencies, they have not destroyed the continents, which still rear their broad plains and great plateaus and mountain ranges above the sea. Either, then, the earth is very young and the agents of denudation have not yet had time to do their work, or they have been opposed successfully by other forces.

We enter now upon a department of our science which treats of forces which work upon the earth from within, and increase the inequalities of its surface. It is they which uplift and recreate the lands which the agents of denudation are continually destroying; it is they which deepen the ocean bed and thus withdraw its waters from the shores. At times also these forces have aided in the destruction of the lands by gradually lowering them and bringing in the sea. Under the action of forces resident within the earth the crust slowly rises or sinks; from time to time it has been folded and broken; while vast quantities of molten rock have been pressed up into it from beneath and outpoured upon its surface. We shall take up these phenomena in the following chapters, which treat of upheavals and depressions of the crust, foldings and fractures of the crust, earthquakes, volcanoes, the interior conditions of the earth, mineral veins, and metamorphism.

OSCILLATIONS OF THE CRUST

Of the various movements of the crust due to internal agencies we will consider first those called oscillations, which lift or depress large areas so slowly that a long time is needed to produce perceptible changes of level, and which leave the strata in nearly their original horizontal attitude. These movements are most conspicuous along coasts, where they can be referred to the datum plane of sea level; we will therefore take our first illustrations from rising and sinking shores.

NEW JERSEY. Along the coasts of New Jersey one may find awash at high tide ancient shell heaps, the remains of tribal feasts of aborigines. Meadows and old forest grounds, with the stumps still standing, are now overflowed by the sea, and fragments of their turf and wood are brought to shore by waves. Assuming that the sea level remains constant, it is clear that the New Jersey coast is now gradually sinking. The rate of submergence has been estimated at about two feet per century.

On the other hand, the wide coastal plain of New Jersey is made of stratified sands and clays, which, as their marine fossils show, were outspread beneath the sea. Their present position above sea level proves that the land now subsiding emerged in the recent past.

The coast of New Jersey is an example of the slow and tranquil oscillations of the earth's unstable crust now in progress along many shores. Some are emerging from the sea, some are sinking beneath it; and no part of the land seems to have been exempt from these changes in the past.

EVIDENCES OF CHANGES OF LEVEL. Taking the surface of the sea as a level of reference, we may accept as proofs of relative upheaval whatever is now found in place above sea level and could have been formed only at or beneath it, and as proofs of relative subsidence whatever is now found beneath the sea and could only have been formed above it.

Thus old strand lines with sea cliffs, wave-cut rock benches, and beaches of wave-worn pebbles or sand, are striking proofs of recent emergence to the amount of their present height above tide. No less conclusive is the presence of sea-laid rocks which we may find in the neighboring quarry or outcrop, although it may have been long ages since they were lifted from the sea to form part of the dry land.

Among common proofs of subsidence are roads and buildings and other works of man, and vegetal growths and deposits, such as forest grounds and peat beds, now submerged beneath the sea. In the deltas of many large rivers, such as the Po, the Nile, the Ganges, and the Mississippi, buried soils prove subsidences of hundreds of feet; and in several cases, as in the Mississippi delta, the depression seems to be now in progress.

Other proofs of the same movement are drowned land forms which are modeled only in open air. Since rivers cannot cut their valleys farther below the baselevel of the sea than the depths of their channels, DROWNED VALLEYS are among the plainest proofs of depression. To this class belong Narragansett, Delaware, Chesapeake, Mobile, and San Francisco bays, and many other similar drowned valleys along the coasts of the United States. Less conspicuous are the SUBMARINE CHANNELS which, as soundings show, extend from the mouths of a number of rivers some distance out to sea. Such is the submerged channel which reaches from New York Bay southeast to the edge of the continental shelf, and which is supposed to have been cut by the Hudson River when this part of the shelf was a coastal plain.

WARPING. In a region undergoing changes of level the rate of movement commonly varies in different parts. Portions of an area may be rising or sinking, while adjacent portions are stationary or moving in the opposite direction. In this way a land surface becomes WARPED. Thus, while Nova Scotia and New Brunswick are now rising from the level of the sea, Prince Edward Island and Cape Breton Island are sinking, and the sea now flows over the site of the famous old town of Louisburg destroyed in 1758.

Since the close of the glacial epoch the coasts of Newfoundland and Labrador have risen hundreds of feet, but the rate of emergence has not been uniform. The old strand line, which stands at five hundred and seventy-five feet above tide at St. John's, Newfoundland, declines to two hundred and fifty feet near the northern point of Labrador.

THE GREAT LAKES is now under-going perceptible warping. Rivers enter the lakes from the south and west with sluggish currents and deep channels resembling the estuaries of drowned rivers; while those that enter from opposite directions are swift and shallow. At the western end of Lake Erie are found submerged caves containing stalactites, and old meadows and forest grounds are now under water. It is thus seen that the water of the lakes is rising along their southwestern shores, while from their north-eastern shores it is being withdrawn. The region of the Great Lakes is therefore warping; it is rising in the northeast as compared with the southwest.

From old bench marks and records of lake levels it has been estimated that the rate of warping amounts to five inches a century for every one hundred miles. It is calculated that the water of Lake Michigan is rising at Chicago at the rate of nine or ten inches per century. The divide at this point between the tributaries of the Mississippi and Lake Michigan is but eight feet above the mean stage of the lake. If the canting of the region continues at its present rate, in a thousand years the waters of the lake will here overflow the divide. In three thousand five hundred years all the lakes except Ontario will discharge by this outlet, via the Illinois and Mississippi rivers, into the Gulf of Mexico. The present outlet by the Niagara River will be left dry, and the divide between the St. Lawrence and the Mississippi systems will have shifted from

Chicago to the vicinity of Buffalo.

PHYSIOGRAPHIC EFFECTS OF OSCILLATIONS. We have already mentioned several of the most important effects of movements of elevation and depression, such as their effects on rivers, the mantle of waste, and the forms of coasts. Movements of elevation--including uplifts by folding and fracture of the crust to be noticed later-- are the necessary conditions for erosion by whatever agent. They determine the various agencies which are to be chiefly concerned m the wear of any land,--whether streams or glaciers, weathering or the wind,--and the degree of their efficiency. The lands must be uplifted before they can be eroded, and since they must be eroded before their waste can be deposited, movements of elevation are a prerequisite condition for sedimentation also. Subsidence is a necessary condition for deposits of great thickness, such as those of the Great Valley of California and the Indo-Gangetic plain (p. 101), the Mississippi delta (p. 109), and the still more important formations of the continental delta in gradually sinking troughs (p. 183). It is not too much to say that the character and thickness of each formation of the stratified rocks depend primarily on these crustal movements.

Along the Baltic coast of Sweden, bench marks show that the sea is withdrawing from the land at a rate which at the north amounts to between three and four feet per century; Towards the south the rate decreases. South of Stockholm, until recent years, the sea has gained upon the land, and here in several seaboard towns streets by the shore are still submerged. The rate of oscillation increases also from the coast inland. On the other hand, along the German coast of the Baltic the only historic fluctuations of sea level are those which may be accounted for by variations due to changes in rainfall. In 1730 Celsius explained the changes of level of the Swedish coast as due to a lowering of the Baltic instead of to an elevation of the land. Are the facts just stated consistent with his theory?

At the little town of Tadousac--where the Saguenay River empties into the St. Lawrence--there are terraces of old sea beaches, some almost as fresh as recent railway fills, the highest standing two hundred and thirty feet above the river. Here the Saguenay is eight hundred and forty feet in depth, and the tide ebbs and flows far up its stream. Was its channel cut to this depth by the river when the land was at its present height? What oscillations are here recorded, and to what amount?

A few miles north of Naples, Italy, the ruins of an ancient Roman temple lie by the edge of the sea, on a narrow plain which is overlooked in the rear by an old sea cliff (Fig. 166). Three marble pillars are still standing. For eleven feet above their bases these columns are uninjured, for to this height they were protected by an accumulation of volcanic ashes; but from eleven to nineteen feet they are closely pitted with the holes of boring marine mollusks. From these facts trace the history of the oscillations of the region.

FOLDINGS OF THE CRUST

The oscillations which we have just described leave the strata not far from their original horizontal attitude. Figure 167 represents a region in which movements of a very different nature have taken place. Here, on either side of the valley V, we find outcrops of layers tilted at high angles. Sections along the ridge r show that it is composed of layers which slant inward from either side. In places the outcropping strata stand nearly on edge, and on the right of the valley they are quite overturned; a shale SH has come to overlie a limestone LM although the shale is the older rock, whose original position was beneath the limestone.

It is not reasonable to suppose that these rocks were deposited in the attitude in which we find them now; we must believe that, like other stratified rocks, they were outspread in nearly level sheets upon the ocean floor. Since that time they must have been deformed. Layers of solid rock several miles in thickness have been crumpled and folded like soft wax in the hand, and a vast denudation has worn away the upper portions of the folds, in part represented in our section by dotted lines.

DIP AND STRIKE. In districts where the strata have been disturbed it is desirable to record their attitude. This is most easily done by taking the angle at which the strata are inclined and the compass direction in which they slant. It is also convenient to record the direction in which the outcrop of the strata trends across the country.

The inclination of a bed of rocks to the horizon is its DIP. The amount of the dip is the angle made with a horizontal plane. The dip of a horizontal layer is zero, and that of a vertical layer is 90 degrees. The direction of the dip is taken with the compass. Thus a geologist's notebook in describing the attitude of outcropping strata contains many such entries as these: dip 32 degrees north, or dip 8 degrees south 20 degrees west,--meaning in the latter case that the amount of the dip is 8 degrees and the direction of the dip bears 20 degrees west of south.

The line of intersection of a layer with the horizontal plane is the STRIKE. The strike always runs at right angles to the dip.

Dip and strike may be illustrated by a book set aslant on a shelf. The dip is the acute angle made with the shelf by the side of the book, while the strike is represented by a line running along the book's upper edge. If the dip is north or south, the strike runs east and west.

FOLDED STRUCTURES. An upfold, in which the strata dip away from a line drawn along the crest and called the axis of the fold, is known as an ANTICLINE. A downfold, where the strata dip from either side toward the axis of the trough, is called a SYNCLINE. There is sometimes seen a downward bend in horizontal or gently inclined strata, by which they descend to a lower level. Such a single flexure is a MONOCLINE.

DEGREES OF FOLDING. Folds vary in degree from broad, low swells, which can hardly be detected, to the most highly contorted and complicated structures. In SYMMETRIC folds the dips of the rocks on each side the axis of the fold are equal. In UNSYMMETRICAL folds one limb is steeper than the other, as in the anticline in Figure 167. In OVERTURNED folds one limb is inclined beyond the perpendicular. FAN FOLDS have been so pinched that the original anticlines are left broader at the top than at the bottom.

In folds where the compression has been great the layers are often found thickened at the crest and thinned along the limbs. Where strong rocks such as heavy limestones are folded together with weak rocks such as shales, the strong rocks are often bent into great simple folds, while the weak rocks are minutely crumpled.

SYSTEMS OF FOLDS. As a rule, folds occur in systems. Over the Appalachian mountain belt, for example, extending from northeastern Pennsylvania to northern Alabama and Georgia, the earth's crust has been thrown into a series of parallel folds whose axes run from northeast to southwest (Fig. 175). In Pennsylvania one may count a score or more of these earth waves,-- some but from ten to twenty miles in length, and some extending as much as two hundred miles before they die away. On the eastern part of this belt the folds are steeper and more numerous than on the western side.

CAUSE AND CONDITIONS OF FOLDING. The sections which we have studied suggest that rocks are folded by lateral pressure. While a single, simple fold might be produced by a heave, a series of folds, including overturns, fan folds, and folds thickened on their crests at the expense of their limbs, could only be made in one way,--by pressure from the side. Experiment has reproduced all forms of folds by subjecting to lateral thrust layers of plastic material such as wax.

Vast as the force must have been which could fold the solid rocks of the crust as one may crumple the leaves of a magazine in the fingers, it is only under certain conditions that it could have produced the results which we see. Rocks are brittle, and it is only when under a HEAVY LOAD and by GREAT PRESSURE SLOWLY APPLIED, that they can thus be folded and bent instead of being crushed to pieces. Under these conditions, experiments prove that not only metals such as steel, but also brittle rocks such as marble, can be deformed and molded and made to flow like plastic clay.

ZONE OF FLOW, ZONE OF FLOW AND FRACTURE, AND ZONE OF FRACTURE. We may believe that at depths which must be reckoned in tens of thousands of feet the load of overlying rocks is so great that rocks of all kinds yield by folding to lateral pressure, and flow instead of breaking. Indeed, at such profound depths and under such inconceivable weight no cavity can form, and any fractures would be healed at once by the welding of grain to grain. At less depths there exists a zone where soft rocks fold and flow under stress, and hard rocks are fractured; while at and near the surface hard and soft rocks alike yield by fracture to strong pressure.

STRUCTURES DEVELOPED IN COMPRESSED ROCKS

Deformed rocks show the effects of the stresses to which they have yielded, not only in the immense folds into which they have been thrown but in their smallest parts as well. A hand specimen of slate, or even a particle under the microscope, may show plications similar in form and origin to the foldings which have produced ranges of mountains. A tiny flake of mica in the rocks of the Alps may be puckered by the same resistless forces which have folded miles of solid rock to form that lofty range.

SLATY CLEAVAGE. Rocks which have yielded to pressure often split easily in a certain direction across the bedding planes. This cleavage is known as slaty cleavage, since it is most perfectly developed in fine-grained, homogeneous rocks, such as slates, which cleave to the thin, smooth-surfaced plates with which we are familiar in the slates used in roofing and for ciphering and blackboards. In coarse-grained rocks, pressure develops more distant partings which separate the rocks into blocks.

Slaty cleavage cannot be due to lamination, since it commonly crosses bedding planes at an angle, while these planes have been often well-nigh or quite obliterated. Examining slate with a microscope, we find that its cleavage is due to the grain of the rock. Its particles are flattened and lie with their broad faces in parallel planes, along which the rock naturally splits more easily than in any other direction. The irregular grains of the mud which has been altered to slate have been squeezed flat by a pressure exerted at right angles to the plane of cleavage. Cleavage is found only in folded rocks, and, as we may see in Figure 176, the strike of the cleavage runs parallel to the strike of the strata and the axis of the folds. The dip of the cleavage is generally steep, hence the pressure was nearly horizontal. The pressure which has acted at right angles to the cleavage, and to which it is due, is the same lateral pressure which has thrown the strata into folds.

We find additional proof that slates have undergone compression at right angles to their cleavage in the fact that any inclusions in them, such as nodules and fossils, have been squeezed out of shape and have their long diameters lying in the planes of cleavage.

That pressure is competent to cause cleavage is shown by experiment. Homogeneous material of fine grain, such as beeswax, when subjected to heavy pressure cleaves at right angles to the direction of the compressing force.

RATE OF FOLDING. All the facts known with regard to rock deformation agree that it is a secular process, taking place so slowly that, like the deepening of valleys by erosion, it escapes the notice of the inhabitants of the region. It is only under stresses slowly applied that rocks bend without breaking. The folds of some of the highest mountains have risen so gradually that strong, well-intrenched rivers which had the right of way across the region were able to hold to their courses, and as a circular saw cuts its way through the log which is steadily driven against it, so these rivers sawed their gorges through the fold as fast as it rose beneath them. Streams which thus maintain the course which they had antecedent to a deformation of the region are known as ANTECEDENT streams. Examples of such are the Sutlej and other rivers of India, whose valleys trench the outer ranges of the Himalayas and whose earlier river deposits have been upturned by the rising ridges. On the other hand, mountain crests are usually divides, parting the head waters of different drainage systems. In these cases the original streams of the region have been broken or destroyed by the uplift of the mountain mass across their paths.

On the whole, which have worked more rapidly, processes of deformation or of denudation?

LAND FORMS DUE TO FOLDING

As folding goes on so slowly, it is never left to form surface features unmodified by the action of other agencies. An anticlinal fold is attacked by erosion as soon as it begins to rise above the original level, and the higher it is uplifted, and the stronger are its slopes, the faster is it worn away. Even while rising, a young upfold is often thus unroofed, and instead of appearing as a long, Smooth, boat-shaped ridge, it commonly has had opened along the rocks of the axis, when these are weak, a valley which is overlooked by the infacing escarpments of the hard layers of the sides of the fold. Under long-continued erosion, anticlines may be degraded to valleys, while the synclines of the same system may be left in relief as ridges.

FOLDED MOUNTAINS. The vastness of the forces which wrinkle the crust is best realized in the presence of some lofty mountain range. All mountains, indeed, are not the result of folding. Some, as we shall see, are due to upwarps or to fractures of the crust; some are piles of volcanic material; some are swellings caused by the intrusion of molten matter beneath the surface; some are the relicts left after the long denudation of high plateaus.

But most of the mountain ranges of the earth, and some of the greatest, such as the Alps and the Himalayas, were originally mountains of folding. The earth's crust has wrinkled into a fold; or into a series of folds, forming a series of parallel ridges and intervening valleys; or a number of folds have been mashed together into a vast upswelling of the crust, in which the layers have been so crumpled and twisted, overturned and crushed, that it is exceedingly difficult to make out the original structure.

The close and intricate folds seen in great mountain ranges were formed, as we have seen, deep below the surface, within the zone of folding. Hence they may never have found expression in any individual surface features. As the result of these deformations deep under ground the surface was broadly lifted to mountain height, and the crumpled and twisted mountain structures are now to be seen only because erosion has swept away the heavy cover of surface rocks under whose load they were developed.

When the structure of mountains has been deciphered it is possible to estimate roughly the amount of horizontal compression which the region has suffered. If the strata of the folds of the Alps were smoothed out, they would occupy a belt seventy-four miles wider than that to which they have been compressed, or twice their present width. A section across the Appalachian folds in Pennyslvania shows a compression to about two thirds the original width; the belt has been shortened thirty-five miles in every hundred.

Considering the thickness of their strata, the compression which mountains have undergone accounts fully for their height, with enough to spare for all that has been lost by denudation.

The Appalachian folds involve strata thirty thousand feet in thickness. Assuming that the folded strata rested on an unyielding foundation, and that what was lost in width was gained in height, what elevation would the range have reached had not denudation worn it as it rose?

THE LIFE HISTORY OF MOUNTAINS. While the disturbance and uplift of mountain masses are due to deformation, their sculpture into ridges and peaks, valleys and deep ravines, and all the forms which meet the eye in mountain scenery, excepting in the very youngest ranges, is due solely to erosion. We may therefore classify mountains according to the degree to which they have been dissected. The Juras are an example of the stage of early youth, in which the anticlines still persist as ridges and the synclines coincide with the valleys; this they owe as much to the slight height of their uplift as to the recency of its date.

The Alps were upheaved at various times, the last uplift being later than the uplift of the Juras, but to so much greater height that erosion has already advanced them well on towards maturity. The mountain mass has been cut to the core, revealing strange contortions of strata which could never have found expression at the surface. Sharp peaks, knife-edged crests, deep valleys with ungraded slopes subject to frequent landslides, are all features of Alpine scenery typical of a mountain range at this stage in its life history. They represent the survival of the hardest rocks and the strongest structures, and the destruction of the weaker in their long struggle for existence against the agents of erosion. Although miles of rock have been removed from such ranges as the Alps, we need not suppose that they ever stood much, if any, higher than at present. All this vast denudation may easily have been accomplished while their slow upheaval was going on; in several mountain ranges we have evidence that elevation has not yet ceased.

Under long denudation mountains are subdued to the forms characteristic of old age. The lofty peaks and jagged crests of their earlier life are smoothed down to low domes and rounded crests. The southern Appalachians and portions of the Hartz Mountains in Germany are examples of mountains which have reached this stage.

There are numerous regions of upland and plains in which the rocks are found to have the same structure that we have seen in folded mountains; they are tilted, crumpled, and overturned, and have clearly suffered intense compression. We may infer that their folds were once lifted to the height of mountains and have since been wasted to low-lying lands. Such a section as that of Figure 67 illustrates how ancient mountains may be leveled to their roots, and represents the final stage to which even the Alps and the Himalayas must sometime arrive. Mountains, perhaps of Alpine height, once stood about Lake Superior; a lofty range once extended from New England and New Jersey southwestward to Georgia along the Piedmont belt. In our study of historic geology we shall see more clearly how short is the life of mountains as the earth counts time, and how great ranges have been lifted, worn away, and again upheaved into a new cycle of erosion.

THE SEDIMENTARY HISTORY OF FOLDED MOUNTAINS. We may mention here some of the conditions which have commonly been antecedent to great foldings of the crust.

1. Mountain ranges are made of belts of enormously and exceptionally thick sediments. The strata of the Appalachians are thirty thousand feet thick, while the same formations thin out to five thousand feet in the Mississippi valley. The folds of the Wasatch Mountains involve strata thirty thousand feet thick, which thin to two thousand feet in the region of the Plains.

2. The sedimentary strata of which mountains are made are for the most part the shallow-water deposits of continental deltas. Mountain ranges have been upfolded along the margins of continents.

3. Shallow-water deposits of the immense thickness found in mountain ranges can be laid only in a gradually sinking area. A profound subsidence, often to be reckoned in tens of thousands of feet, precedes the upfolding of a mountain range.

Thus the history of mountains of folding is as follows: For long ages the sea bottom off the coast of a continent slowly subsides, and the great trough, as fast as it forms, is filled with sediments, which at last come to be many thousands of feet thick. The downward movement finally ceases. A slow but resistless pressure sets in, and gradually, and with a long series of many intermittent movements, the vast mass of accumulated sediments is crumpled and uplifted into a mountain range.

FRACTURES AND DISLOCATIONS OF THE CRUST

Considering the immense stresses to which the rocks of the crust are subjected, it is not surprising to find that they often yield by fracture, like brittle bodies, instead of by folding and flowing, like plastic solids. Whether rocks bend or break depends on the character and condition of the rocks, the load of overlying rocks which they bear, and the amount of the force and the slowness with which it is applied.

JOINTS. At the surface, where their load is least, we find rocks universally broken into blocks of greater or less size by partings known as joints. Under this name are included many division planes caused by cooling and drying; but it is now generally believed that the larger and more regular joints, especially those which run parallel to the dip and strike of the strata, are fractures due to up-and-down movements and foldings and twistings of the rocks.

Joints are used to great advantage in quarrying, and we have seen how they are utilized by the weather in breaking up rock masses, by rivers in widening their valleys, by the sea in driving back its cliffs, by glaciers in plucking their beds, and how they are enlarged in soluble rocks to form natural passageways for underground waters. The ends of the parted strata match along both sides of joint planes; in. joints there has been little or no displacement of the broken rocks.

FAULTS. In Figure 184 the rocks have been both broken and dislocated along the plane ff. One side must have been moved up or down past the other. Such a dislocation is called a fault. The amount of the displacement, as measured by the vertical distance between the ends of a parted layer, is the throw. The angle which the fault plane makes with the vertical is the HADE. In Figure 184 the right side has gone down relatively to the left; the right is the side of the downthrow, while the left is the side of the upthrow. Where the fault plane is not vertical the surfaces on the two sides may be distinguished as the HANGING WALL and the FOOT WALL. Faults differ in throw from a fraction of an inch to many thousands of feet.

SLICKENSIDES. If we examine the walls of a fault, we may find further evidence of movement in the fact that the surfaces are polished and grooved by the enormous friction which they have suffered as they have ground one upon the other. These appearances, called sliekensides, have sometimes been mistaken for the results of glacial action.

NORMAL FAULTS. Faults are of two kinds,--normal faults and thrust faults. Normal faults, of which Figure 184 is an example, hade to the downthrow; the hanging wall has gone down. The total length of the strata has been increased by the displacement. It seems that the strata have been stretched and broken, and that the blocks have readjusted themselves under the action of gravity as they settled.

THRUST FAULTS. Thrust faults hade to the upthrow; the hanging wall has gone up. Clearly such faults, where the strata occupy less space than before, are due to lateral thrust. Folds and thrust faults are closely associated. Under lateral pressure strata may fold to a certain point and then tear apart and fault along the surface of least resistance. Under immense pressure strata also break by shear without folding. Thus, in Figure 185, the rigid earth block under lateral thrust has found it easier to break along the fault plane than to fold. Where such faults are nearly horizontal they are distinguished as THRUST PLANES.

In all thrust faults one mass has been pushed over another, so as to bring the underlying and older strata upon younger beds; and when the fault planes are nearly horizontal, and especially when the rocks have been broken into many slices which have slidden far one upon another, the true succession of strata is extremely hard to decipher.

In the Selkirk Mountains of Canada the basement rocks of the region have been driven east for seven miles on a thrust plane, over rocks which originally lay thousands of feet above them.

Along the western Appalachians, from Virginia to Georgia, the mountain folds are broken by more than fifteen parallel thrust planes, running from northeast to southwest, along which the older strata have been pushed westward over the younger. The longest continuous fault has been traced three hundred and seventy-five miles, and the greatest horizontal displacement has been estimated at not less than eleven miles.

CRUSH BRECCIA. Rocks often do not fault with a clean and simple fracture, but along a zone, sometimes several yards in width, in which they are broken to fragments. It may occur also that strata which as a whole yield to lateral thrust by folding include beds of brittle rocks, such as thin-layered limestones, which are crushed to pieces by the strain. In either case the fragments when recemented by percolating waters form a rock known as a CRUSH BRECCIA (pronounced BRETCHA).

Breccia is a term applied to any rock formed of cemented ANGULAR fragments. This rock may be made by the consolidation of volcanic cinders, of angular waste at the foot of cliffs, or of fragments of coral torn by the waves from coral reefs, as well as of strata crushed by crustal movements.

SURFACE FEATURES DUE TO DISLOCATIONS

FAULT SCARPS. A fault of recent date may be marked at surface by a scarp, because the face of the upthrown block has not yet been worn to the level of the downthrow side.

After the upthrown block has been worn down to this level, differential erosion produces fault scarps wherever weak rocks and resistant rocks are brought in contact along the fault plane; and the harder rocks, whether on the upthrow or the downthrow side, emerge in a line of cliffs. Where a fault is so old that no abrupt scarps appear, its general course is sometimes marked by the line of division between highland and lowland or hill and plain. Great faults have sometimes brought ancient crystalline rocks in contact with weaker and younger sedimentary rocks, and long after erosion has destroyed all fault scarps the harder crystallines rise in an upland of rugged or mountainous country which meets the lowland along the line of faulting.

The vast majority of faults give rise to no surface features. The faulted region may be old enough to have been baseleveled, or the rocks on both sides of the line of dislocation may be alike in their resistance to erosion and therefore have been worn down to a common slope. The fault may be entirely concealed by the mantle of waste, and in such cases it can be inferred from abrupt changes in the character or the strike and dip of the strata where they may outcrop near it.

The plateau trenched by the Grand Canyon of the Colorado River exhibits a series of magnificent fault scarps whose general course is from north to south, marking the edges of the great crust blocks into which the country has been broken. The highest part of the plateau is a crust block ninety miles long and thirty-five miles in maximum width, which has been hoisted to nine thousand three hundred feet above, sea level. On the east it descends four thousand feet by a monoclinal fold, which passes into a fault towards the north. On the west it breaks down by a succession of terraces faced by fault scarps. The throw of these faults varies from seven hundred feet to more than a mile. The escarpments, however, are due in a large degree to the erosion of weaker rock on the downthrow side.

The Highlands of Scotland meet the Lowlands on the south with a bold front of rugged hills along a line of dislocation which runs across the country from sea to sea. On the one side are hills of ancient crystalline rocks whose crumpled structures prove that they are but the roots of once lofty mountains; on the other lies a lowland of sandstone and other stratified rocks formed from the waste of those long-vanished mountain ranges. Remnants of sandstone occur in places on the north of the great fault, and are here seen to rest on the worn and fairly even surface of the crystallines. We may infer that these ancient mountains were reduced along their margins to low plains, which were slowly lowered beneath the sea to receive a cover of sedimentary rocks. Still later came an uplift and dislocation. On the one side erosion has since stripped off the sandstones for the most part, but the hard crystalline rocks yet stand in bold relief. On the other side the weak sedimentary rocks have been worn down to lowlands.

RIFT VALLEYS. In a broken region undergoing uplift or the unequal settling which may follow, a slice inclosed between two fissures may sink below the level of the crust blocks on either side, thus forming a linear depression known as a rift valley, or valley of fracture.

One of the most striking examples of this rare type of valley is the long trough which runs straight from the Lebanon Mountains of Syria on the north to the Red Sea on the south, and whose central portion is occupied by the Jordan valley and the Dead Sea. The plateau which it gashes has been lifted more than three thousand feet above sea level, and the bottom of the trough reaches a depth of two thousand six hundred feet below that level in parts of the Dead Sea. South of the Dead Sea the floor of the trough rises somewhat above sea level, and in the Gulf of Akabah again sinks below it. This uneven floor could be accounted for either by the profound warping of a valley of erosion or by the unequal depression of the floor of a rift valley. But that the trough is a true valley of fracture is proved by the fact that on either side it is bounded by fault scarps and monoclinal folds. The keystone of the arch has subsided. Many geologists believe that the Jordan- Akabah trough, the long narrow basin of the Red Sea, and the chain of down-faulted valleys which in Africa extends from the strait of Bab-el-Mandeb as far south as Lake Nyassa--

valleys which contain more than thirty lakes--belong to a single system of dislocation.

Should you expect the lateral valleys of a rift valley at the time of its formation to enter it as hanging valleys or at a common level?

BLOCK MOUNTAINS. Dislocations take place on so grand a scale that by the upheaval of blocks of the earth's crust or the down- faulting of the blocks about one which is relatively stationary, mountains known as block mountains are produced. A tilted crust block may present a steep slope on the side upheaved and a more gentle descent on the side depressed.

THE BASIN RANGES. The plateaus of the United States bounded by the Rocky Mouirtains on the east, and on the west by the ranges which front the Pacific, have been profoundly fractured and faulted. The system of great fissures by which they are broken extends north and south, and the long, narrow, tilted crust blocks intercepted between the fissures give rise to the numerous north-south ranges of the region. Some of the tilted blocks, as those of southern Oregon, are as yet but moderately carved by erosion, and shallow lakes lie on the waste that has been washed into the depressions between them. We may therefore conclude that their displacement is somewhat recent. Others, as those of Nevada, are so old that they have been deeply dissected; their original form has been destroyed by erosion, and the intermontane depressions are occupied by wide plains of waste.

DISLOCATIONS AND RIVER VALLEYS. Before geologists had proved that rivers can by their own unaided efforts cut deep canyons, it was common to consider any narrow gorge as a gaping fissure of the crust. This crude view has long since been set aside. A map of the plateaus of northern Arizona shows how independent of the immense faults of the region is the course of the Colorado River. In the Alps the tunnels on the Saint Gotthard railway pass six times beneath the gorge of the Reuss, but at no point do the rocks show the slightest trace of a fault.

RATE OF DISLOCATION. So far as human experience goes, the earth movements which we have just studied, some of which have produced deep-sunk valleys and lofty mountain ranges, and faults whose throw is to be measured in thousands of feet, are slow and gradual. They are not accomplished by a single paroxysmal effort, but by slow creep and a series of slight slips continued for vast lengths of time.

In the Aspen mining district in Colorado faulting is now going on at a comparatively rapid rate. Although no sudden slips take place, the creep of the rock along certain planes of faulting gradually bends out of shape the square-set timbers in horizontal drifts and has closed some vertical shafts by shifting the upper portion across the lower. Along one of the faults of this region it is estimated that there has been a movement of at least four hundred feet since the Glacial epoch. More conspicuous are the instances of active faulting by means of sudden slips. In 1891 there occurred along an old fault plane in Japan a slip which produced an earth rent traced for fifty miles (Fig. 192). The country on one side was depressed in places twenty feet below that on the other, and also shifted as much as thirteen feet horizontally in the direction of the fault line.

In 1872 a slip occurred for forty miles on the great line of dislocation which runs along the eastern base of the Sierra Nevada Mountains. In the Owens valley, California, the throw amounted to twenty-five feet in places, with a horizontal movement along the fault line of as much as eighteen feet. Both this slip and that in Japan just mentioned caused severe earthquakes.

For the sake of clearness we have described oscillations, foldings, and fractures of the crust as separate processes, each giving rise to its own peculiar surface features, but in nature earth movements are by no means so simple,-- they are often implicated with one another: folds pass into faults; in a deformed region certain rocks have bent, while others under the same strain, but under different conditions of plasticity and load, have broken; folded mountains have been worn to their roots, and the peneplains to which they have been denuded have been upwarped to mountain height and afterwards dissected,--as in the case of the Alleghany ridges, the southern Carpathians, and other ranges, --or, as in the case of the Sierra Nevada Mountains, have been broken and uplifted as mountains of fracture.

Draw the following diagrams, being careful to show the direction in which the faulted blocks have moved, by the position of the two parts of some well-defined layer of limestone, sandstone, or shale, which occurs on each side of the fault plane, as in Figure 184.

1. A normal fault with a hade of 15 degrees, the original fault scarp remaining.

2. A normal fault with a hade of 50 degrees, the original fault scarp worn away, showing cliffs caused by harder strata on the downthrow side.

3. A thrust fault with a hade of 30 degrees, showing cliffs due to harder strata outcropping on the downthrow.

4. A thrust fault with a hade of 80 degrees, with surface baseleveled.

5. In a region of normal faults a coal mine is being worked along the seam of coal AB (Fig. 193). At B it is found broken by a fault f which hades toward A. To find the seam again, should you advise tunneling up or down from B?

6. In a vertical shaft of a coal mine the same bed of coal is pierced twice at different levels because of a fault. Draw a diagram to show whether the fault is normal or a thrust.

7. Copy the diagram in Figure 194, showing how the two ridges may be accounted for by a single resistant stratum dislocated by a fault. Is the fault a STRIKE FAULT, i.e. one running parallel with the strike of the strata, or a DIP FAULT, one running parallel with the direction of the dip?

8. Draw a diagram of the block in Figure 195 as it would appear if dislocated along the plane efg by a normal fault whose throw equals one fourth the height of the block. Is the fault a strike or a dip fault? Draw a second diagram showing the same block after denudation has worn it down below the center of the upthrown side. Note that the outcrop of the coal seam is now deceptively repeated. This exercise may be done in blocks of wood instead of drawings.

9. Draw diagrams showing by dotted lines the conditions both of A and of B, Figure 196, after deformation had given the strata their present attitude.

10. What is the attitude of the strata of this earth block, Figure 197? What has taken place along the plane bef? When did the dislocation occur compared with the folding of the strata? With the erosion of the valleys on the right-hand

side of the mountain? With the deposition of the sediments? Do you find any remnants of the original surface baf produced by the dislocation? From the left-hand side of the mountain infer what was the relief of the region before the dislocation. Give the complete history recorded in the diagram from the deposition of the strata to the present.

11. Which is the older fault, in Figure 198, or When did the lava flow occur? How long a time elapsed between the formation of the two faults as measured in the work done in the interval? How long a time since the formation of the later fault?

12. Measure by the scale the thickness lie of the coal-bearing strata outcropping from a to b in Figure 199. On any convenient scale draw a similar section of strata with a dip of 30 degrees outcropping along a horizontal line normal to the strike one thousand feet in length, and measure the thickness of the strata by the scale employed. The thickness may also be calculated by trigonometry.

UNCONFORMITY

Strata deposited one upon, another in an unbroken succession are said to be conformable. But the continuous deposition of strata is often interrupted by movements of the earth's crust, Old sea floors are lifted to form land and are again depressed beneath the sea to receive a cover of sediments only after an interval during which they were carved by subaerial erosion. An erosion surface which thus parts older from younger strata is known as an UNCONFORMITY, and the strata above it are said to be UNCONFORMABLE with the rocks below, or to rest unconformably upon them. An unconformity thus records movements of the crust and a consequent break in the deposition of the strata. It denotes a period of land erosion of greater or less length, which may sometimes be roughly measured by the stage in the erosion cycle which the land surface had attained before its burial. Unconformable strata may be parallel, as in Figure 200, where the record includes the deposition of strata, their emergence, the erosion of the land surface, a submergence and the deposit of the strata, and lastly, emergence and the erosion of the present surface.

Often the earth movements to which the uplift or depression was due involved tilting or folding of the earlier strata, so that the strata are now nonparallel as well as unconformable. In Figure 201, for example, the record includes deposition, uplift, and tilting of a; erosion, depression, the deposit of b; and finally the uplift which has brought the rocks to open air and permitted the dissection by which the unconformity is revealed. From this section infer that during early Silurian times the area was sea, and thick sea muds were laid upon it. These were later altered to hard slates by pressure and upfolded into mountains. During the later Silurian and the Devonian the area was land and suffered vast denudation. In the Carboniferous period it was lowered beneath the sea and received a cover of limestone.

THE AGE OF MOUNTAINS. It is largely by means of unconformities that we read the history of mountain making and other deformations and movements of the crust. In Figure 203, for example, the deformation which upfolded the range of mountains took place after the deposit of the series of strata a of which the mountains are composed, and before the deposit of the stratified rocks, which rest unconformably on a and have not shared their uplift.

Most great mountain ranges, like the Sierra Nevada and the Alps, mark lines of weakness along which the earth's crust has yielded again and again during the long ages of geological time. The strata deposited at various times about their flanks have been infolded by later crumplings with the original mountain mass, and have been repeatedly crushed, inverted, faulted, intruded with igneous rocks, and denuded. The structure of great mountain ranges thus becomes exceedingly complex and difficult to read. A comparatively simple case of repeated uplift is shown in Figure 204. In the section of a portion of the Alps shown in Figure 179 a far more complicated history may be deciphered.

UNCONFORMITIES IN THE COLORADO CANYON, ARIZONA. How geological history may be read in unconformities is further illustrated in Figures 207 and 208. The dark crystalline rocks a at the bottom of the canyon are among the most ancient known, and are overlain unconformably by a mass of tilted coarse marine sandstones b, whose total thickness is not seen in the diagram and measures twelve thousand feet perpendicularly to the dip. Both a and b rise to a common level nn and upon them rest the horizontal sea-laid strata c, in which the upper portion of the canyon has been cut.

Note that the crystalline rocks a have been crumpled and crushed. Comparing their structure with that of folded mountains, what do you infer as to their relief after their deformation? To which surface were they first worn down, mm' or nn? Describe and account for the surface mm'. How does it differ from the surface of the crystalline rocks seen in the Torridonian Mountains, and why? This surface mm' is one of the oldest land surfaces of which any vestige remains.

It is a bit of fossil geography buried from view since the earliest geological ages and recently brought to light by the erosion of the canyon.

How did the surface mm' come to receive its cover of sandstones b? From the thickness and coarseness of these sediments draw inferences as to the land mass from which they were derived. Was it rising or subsiding? high or low? Were its streams slow or swift? Was the amount of erosion small or great?

Note the strong dip of these sandstones b. Was the surface mm' tilted as now when the sandstones were deposited upon it? When was it tilted? Draw a diagram showing the attitude of the rocks after this tilting occurred, and their height relative to sea level.

The surface nn' is remarkably even, although diversified by some low hills which rise into the bedded rocks of c, and it may be traced for long distances up and down the canyon. Were the layers of b and the surface mm' always thus cut short by nn' as now? What has made the surface nn' so even? How does it come to cross the hard crystalline rocks a and the weaker sandstones b at the same impartial level? How did the sediments of c come to be laid upon it? Give now the entire history recorded in the section, and in addition that involved in the production of the platform P, shown in Figure 130, and that of the cutting of the canyon. How does the time involved in the cutting of the canyon compare with that required for the production of the surfaces mm', nn', and P?

CHAPTER X

EARTHQUAKES

Any sudden movement of the rocks of the crust, as when they tear apart when a fissure is formed or extended, or slip from time to time along a growing fault, produces a jar called an earthquake, which spreads in all directions from the place of disturbance.

THE CHARLESTON EARTHQUAKE. On the evening of August 31, 1886, the city of Charleston, S.C., was shaken by one of the greatest earthquakes which has occurred in the United States. A slight tremor which rattled the windows was followed a few seconds later by a roar, as of subterranean thunder, as the main shock passed beneath the city. Houses swayed to and fro, and their heaving floors overturned furniture and threw persons off their feet as, dizzy and nauseated, they rushed to the doors for safety. In sixty seconds a number of houses were completely wrecked, fourteen thousand chimneys were toppled over, and in all the city scarcely a building was left without serious injury. In the vicinity of Charleston railways were twisted and trains derailed. Fissures opened in the loose superficial deposits, and in places spouted water mingled with sand from shallow underlying aquifers.

The point of origin, or FOCUS, of the earthquake was inferred from subsequent investigations to be a rent in the rocks about twelve miles beneath the surface. From the center of greatest disturbance, which lay above the focus, a few miles northwest of the city, the surface shock traveled outward in every direction, with decreasing effects, at the rate of nearly two hundred miles per minute. It was felt from Boston to Cuba, and from eastern Iowa to the Bermudas, over a circular area whose diameter was a thousand miles.

An earthquake is transmitted from the focus through the elastic rocks of the crust, as a wave, or series of waves, of compression and rarefaction, much as a sound wave is transmitted through the elastic medium of the air. Each earth particle vibrates with exceeding swiftness, but over a very short path. The swing of a particle in firm rock seldom exceeds one tenth of an inch in ordinary earthquakes, and when it reaches one half an inch and an inch, the movement becomes dangerous and destructive.

The velocity of earthquake waves, like that of all elastic waves, varies with the temperature and elasticity of the medium. In the deep, hot, elastic rocks they speed faster than in the cold and broken rocks near the surface. The deeper the point of origin and the more violent the initial shock, the faster and farther do the vibrations run.

Great earthquakes, caused by some sudden displacement or some violent rending of the rocks, shake the entire planet. Their waves run through the body of the earth at the rate of about three hundred and fifty miles a minute, and more slowly round its circumference, registering their arrival at opposite sides of the globe on the exceedingly delicate instruments of modern earthquake observatories.

GEOLOGICAL EFFECTS. Even great earthquakes seldom produce geological effects of much importance. Landslides may be shaken down from the sides of mountains and hills, and cracks may be opened in the surface deposits of plains; but the transient shiver, which may overturn cities and destroy thousands of human lives, runs through the crust and leaves it much the same as before.

EARTHQUAKES ATTENDING GREAT DISPLACEMENTS. Great earthquakes frequently attend the displacement of large masses of the rocks of the crust. In 1822 the coast of Chile was suddenly raised three or four feet, and the rise was five or six feet a mile inland. In 1835 the same region was again upheaved from two to ten feet. In each instance a destructive earthquake was felt for one thousand miles along the coast.

THE GREAT CALIFORNIA EARTHQUAKE OF 1906. A sudden dislocation occurred in 1906 along an ancient fault plane which extends for 300 miles through western California. The vertical displacement did not exceed four feet, while the horizontal shifting reached a maximum of twenty feet. Fences, rows of trees, and roads which crossed the fault were broken and offset. The latitude and longitude of all points over thousands of square miles were changed. On each side of the fault the earth blocks moved in opposite directions, the block on the east moving southward and that on the west moving northward and to twice the distance. East and west of the fault the movements lessened with increasing distance from it.

This sudden slip set up an earthquake lasting sixty-five seconds, followed by minor shocks recurring for many days. In places the jar shook down the waste on steep hillsides, snapped off or uprooted trees, and rocked houses from their foundations or threw down their walls or chimneys. The water mains of San Francisco were broken, and the city was thus left defenseless against a conflagration which destroyed $500,000,000 worth of property. The destructive effects varied with the nature of the ground. Buildings on firm rock suffered least, while those on deep alluvium were severely shaken by the undulations, like water waves, into which the loose material was thrown. Well-braced steel structures, even of the largest size, were earthquake proof, and buildings of other materials, when honestly built and intelligently designed to withstand earthquake shocks, usually suffered little injury. The length of the intervals between severe earthquakes in western California shows that a great dislocation so relieves the stresses of the adjacent earth blocks that scores of years may elapse before the stresses again accumulate and cause another dislocation.

Perhaps the most violent earthquake which ever visited the United States attended the depression, in 1812, of a region seventy-five miles long and thirty miles wide, near New Madrid, Mo. Much of the area was converted into swamps and some into shallow lakes, while a region twenty miles in diameter was bulged up athwart the channel of the Mississippi. Slight quakes are still felt in this region from time to time, showing that the strains to which the dislocation was due have not yet been fully relieved.

EARTHQUAKES ORIGINATING BENEATH THE SEA. Many earthquakes originate beneath the sea, and in a number of examples they seem to have been accompanied, as soundings indicate, by local subsidences of the ocean bottom. There have been instances where the displacement has been sufficient to set the entire Pacific Ocean pulsating for many hours. In mid ocean the wave thus produced has a height of only a few feet, while it may be two hundred miles in width. On shores near the point of origin destructive waves two or three score feet in height roll in, and on coasts thousands of miles distant the expiring undulations may be still able to record themselves on tidal gauges.

DISTRIBUTION OF EARTHQUAKES. Every half hour some considerable area of the earth's surface is sensibly shaken by an earthquake, but earthquakes are by no means uniformly distributed over the globe. As we might infer from what we know as to their causes, earthquakes are most frequent in regions now undergoing deformation. Such are young rising mountain ranges, fault lines where readjustments recur from time to time, and the slopes of suboceanic depressions whose steepness suggests that subsidence may there be in progress.

Earthquakes, often of extreme severity, frequently visit the lofty and young ranges of the Andes, while they are little known in the subdued old mountains of Brazil. The Highlands of Scotland are crossed by a deep and singularly straight depression called the Great Glen, which has been excavated along a very ancient line of dislocation. The earthquakes which occur from time to time in this region, such as the Inverness earthquake in 1891, are referred to slight slips along this fault plane.

In Japan, earthquakes are very frequent. More than a thousand are recorded every year, and twenty-nine world-shaking earthquakes occurred in the three years ending with 1901. They originate, for the most part, well down on the eastern flank of the earth fold whose summit is the mountainous crest of the islands, and which plunges steeply beneath the sea to the abyss of the Tuscarora Deep.

MINOR CAUSES OF EARTHQUAKES. Since any concussion within the crust sets up an earth jar, there are several minor causes of earthquakes, such as volcanic explosions and even the collapse of the roofs of caves. The earthquakes which attend the eruption of volcanoes are local, even in the case of the most violent volcanic paroxysms known. When the top of a volcano has been blown to fragments, the accompanying earth shock has sometimes not been felt more than twenty-five miles away.

DEPTH OF FOCUS. The focus of the Charleston earthquake, estimated at about twelve miles below the surface, was exceptionally deep. Volcanic earthquakes are particularly shallow, and probably no earthquakes known have started at a greater depth than fifteen or twenty miles. This distance is so slight compared with the earth's radius that we may say that earthquakes are but skin-deep.

Should you expect the velocity of an earthquake to be greater in a peneplain or in a river delta?

After an earthquake, piles on which buildings rested were found driven into the ground, and chimneys crushed at base. From what direction did the shock come?

Chimneys standing on the south walls of houses toppled over on the roof.

Should you infer that the shock in this case came from the north or south?

How should you expect a shock from the east to affect pictures hanging on the east and the west walls of a room? how the pictures hanging on the north and the south walls?

In parts of the country, as in southwestern Wisconsin, slender erosion pillars, or "monuments," are common. What inference could you draw as to the occurrence in such regions of severe earthquakes in the recent past?

CHAPTER XI

VOLCANOES

Connected with movements of the earth's crust which take place so slowly that they can be inferred only from their effects is one of the most rapid and impressive of all geological processes,--the extrusion of molten rock from beneath the surface of the earth, giving rise to all the various phenomena of volcanoes.

In a volcano, molten rock from a region deep below, which we may call its reservoir, ascends through a pipe or fissure to the surface. The materials erupted may be spread over vast areas, or, as is commonly the case, may accumulate about the opening, forming a conical pile known as the volcanic cone. It is to this cone that popular usage refers the word VOLCANO; but the cone is simply a conspicuous part of the volcanic mechanism whose still more important parts, the reservoir and the pipe, are hidden from view.

Volcanic eruptions are of two types,--EFFUSIVE eruptions, in which molten rock wells up from below and flows forth in streams of LAVA (a comprehensive term applied to all kinds of rock emitted from volcanoes in a molten state), and EXPLOSIVE eruptions, in which the rock is blown out in fragments great and small by the expansive force of steam.

ERUPTIONS OF THE EFFUSIVE TYPE

THE HAWAIIAN VOLCANOES. The Hawaiian Islands are all volcanic in origin, and have a linear arrangement characteristic of many volcanic groups in all parts of the world. They are strung along a northwest-southeast line, their volcanoes standing in two parallel rows as if reared along two adjacent lines of fracture or folding. In the northwestern islands the volcanoes have long been extinct and are worn low by erosion. In the southeastern island. Hawaii, three volcanoes are still active and in process of building. Of these Mauna Loa, the monarch of volcanoes, with a girth of two hundred miles and a height of nearly fourteen thousand feet above sea level, is a lava dome the slope of whose sides does not average more than five degrees. On the summit is an elliptical basin ten miles in circumference and several hundred feet deep. Concentric cracks surround the rim, and from time to time the basin is enlarged as great slices are detached from the vertical walls and engulfed.

Such a volcanic basin, formed by the insinking of the top of the cone, is called a CALDERA.

On the flanks of Mauna Loa, four thousand feet above sea level, lies the caldera of Kilauea, an independent volcano whose dome has been joined to the larger mountain by the gradual growth of the two. In each caldera the floor, which to the eye is a plain of black lava, is the congealed surface of a column of molten rock. At times of an eruption lakes of boiling lava appear which may be compared to air holes in a frozen river. Great waves surge up, lifting tons of the fiery liquid a score of feet in air, to fall back with a mighty plunge and roar, and occasionally the lava rises several hundred feet in fountains of dazzling brightness. The lava lakes may flood the floor of the basin, but in historic times have never been known to fill it and overflow the rim. Instead, the heavy column of lava breaks way through the sides of the mountain and discharges in streams which flow down the mountain slopes for a distance sometimes of as much as thirty-five miles. With the drawing off of the lava the column in the duct of the volcano lowers, and the floor of the caldera wholly or in part subsides. A black and steaming abyss marks the place of the lava lakes. After a time the lava rises in the duct, the floor is floated higher, and the boiling lakes reappear.

The eruptions of the Hawaiian volcanoes are thus of the effusive type. The column of lava rises, breaks through the side of the mountain, and discharges in lava streams. There are no explosions, and usually no earthquakes, or very slight ones, accompany the eruptions. The lava in the calderas boils because of escaping steam, but the vapor emitted is comparatively little, and seldom hangs above the summits in heavy clouds. We see here in its simplest form the most impressive and important fact in all volcanic action, molten rock has been driven upward to the surface from some deep-lying source.

LAVA FLOWS. As lava issues from the side of a volcano or overflows from the summit, it flows away in a glowing stream resembling molten iron drawn white-hot from an iron furnace. The surface of the stream soon cools and blackens, and the hard crust of nonconducting rock may grow thick and firm enough to form a tunnel, within which the fluid lava may flow far before it loses its heat to any marked degree. Such tunnels may at last be left as caves by the draining away of the lava, and are sometimes several miles in length.

PAHOEHOE AND AA. When the crust of highly fluid lava remains unbroken after its first freezing, it presents a smooth, hummocky, and ropy surface known by the Hawaiian term PAHOEHOE. On the other hand, the crust of a viscid flow may be broken and splintered as it is dragged along by the slowly moving mass beneath. The stream then appears as a field of stones clanking and grinding on, with here and there from some chink a dull red glow or a wisp of steam. It sets to a surface called AA, of broken, sharp-edged blocks, which is often both difficult and dangerous to traverse.

FISSURE ERUPTIONS. Some of the largest and most important outflows of lava have not been connected with volcanic cones, but have been discharged from fissures, flooding the country far and wide with molten rock. Sheet after sheet of molten rock has been successively outpoured, and there have been built up, layer upon layer, plateaus of lava thousands of feet in thickness and many thousands of square miles in area.

ICELAND. This island plateau has been rent from time to time by fissures from which floods of lava have outpoured. In some instances the lava discharges along the whole length of the fissure, but more often only at certain points upon it. The Laki fissure, twenty miles long, was in eruption in 1783 for seven months. The inundation of fluid rock which poured from it is the largest of historic record, reaching a distance of forty-seven miles and covering two hundred and twenty square miles to an average depth of a hundred feet. At the present time the fissure is traced by a line of several hundred insignificant mounds of fragmental materials which mark where the lava issued.

The distance to which the fissure eruptions of Iceland flow on slopes extremely gentle is noteworthy. One such stream is ninety miles in length, and another seventy miles long has a slope of little more than one half a degree.

Where lava is emitted at one point and flows to a less distance there is gradually built up a dome of the shape of an inverted saucer with an immense base but comparatively low. Many LAVA DOMES have been discovered in Iceland, although from their exceedingly gentle slopes, often but two or three degrees, they long escaped the notice of explorers.

The entire plateau of Iceland, a region as large as Ohio, is composed of volcanic products,--for the most part of successive sheets of lava whose total thickness falls little short of two miles. The lava sheets exposed to view were outpoured in open air and not beneath the sea; for peat bogs and old forest grounds are interbedded with them, and the fossil plants of these vegetable deposits prove that the plateau has long been building and is very ancient. On the steep sea cliffs of the island, where its structure is exhibited, the sheets of lava are seen to be cut with many DIKES,--fissures which have been filled by molten rock,--and there is little doubt that it was through these fissures that the lava outwelled in successive flows which spread far and wide over the country and gradually reared the enormous pile of the plateau.

ERUPTIONS OF THE EXPLOSIVE TYPE

In the majority of volcanoes the lava which rises in the pipe is at least in part blown into fragments with violent explosions and shot into the air together with vast quantities of water vapor and various gases. The finer particles into-- which the lava is exploded are called VOLCANIC DUST or VOLCANIC ASHES, and are often carried long distances by the wind before they settle to the earth. The coarser fragments fall about the vent and there accumulate in a steep, conical, volcanic mountain. As successive explosions keep open the throat of the pipe, there remains on the summit a cup-shaped depression called the CRATER.

STROMBOLI. To study the nature of these explosions we may visit Stromboli, a low volcano built chiefly of fragmental materials, which rises from the sea off the north coast of Sicily and is in constant though moderate action.

Over the summit hangs a cloud of vapor which strikingly resembles the column of smoke puffed from the smokestack of a locomotive, in that it consists of globular masses, each the product of a distinct explosion. At night the cloud of vapor is lighted with a red glow at intervals of a few minutes, like the glow on the trail of smoke behind the locomotive when from time to time the fire bos is opened. Because of this intermittent light flashing thousands of feet above the sea, Stromboli has been given the name of the Lighthouse of the Mediterranean.

Looking down into the crater of the volcano, one sees a viscid lava slowly seething. The agitation gradually increases. A great bubble forms. It bursts with an explosion which causes the walls of the crater to quiver with a miniature earthquake, and an outrush of steam carries the fragments of the bubble aloft for a thousand feet to fall into the crater or on the mountain side about it. With the explosion the cooled and darkened crust of the lava is removed, and the light of the incandescent liquid beneath is reflected from the cloud of vapor which overhangs the cone.

At Stromboli we learn the lesson that the explosive force in volcanoes is that of steam. The lava in the pipe is permeated with it much as is a thick boiling porridge. The steam in boiling porridge is unable to escape freely and gathers into bubbles which in breaking spurt out drops of the pasty substance; in the same way the explosion of great bubbles of steam in the viscid lava shoots clots and fragments of it into the air.

KRAKATOA. The most violent eruption of history, that of Krakatoa, a small volcanic island in the strait between Sumatra and Java, occurred in the last week of August, 1883. Continuous explosions shot a column of steam and ashes. seventeen miles in air. A black cloud, beneath which was midnight darkness and from which fell a rain of ashes and stones, overspread the surrounding region to a distance of one hundred and fifty miles. Launched on the currents of the upper air, the dust was swiftly carried westward to long distances. Three days after the eruption it fell on the deck of a ship sixteen hundred miles away, and in thirteen days the finest impalpable powder from the volcano had floated round the globe. For many months the dust hung over Europe and America as a faint lofty haze illuminated at sunrise and sunset with brilliant crimson. In countries nearer the eruption, as in India and Africa, the haze for some time was so thick that it colored sun and moon with blue, green, and copper-red tints and encircled them with coronas.

At a distance of even a thousand miles the detonations of the eruption sounded like the booming of heavy guns a few miles away. In one direction they were audible for a distance as great as that from San Francisco to Cleveland. The entire atmosphere was thrown into undulations under which all barometers rose and fell as the air waves thrice encircled the earth. The shock of the explosions raised sea waves which swept round the adjacent shores at a height of more than fifty feet, and which were perceptible halfway around the globe.

At the close of the eruption it was found that half the mountain had been blown away, and that where the central part of the island had been the sea was a thousand feet deep.

MARTINIQUE AND ST. VINCENT. In 1902 two dormant volcanoes of the West Indies, Mt. Pelee in Martinique and Soufriere in St. Vincent, broke into eruption simultaneously. No lava was emitted, but there were blown into the air great quantities of ashes, which mantled the adjacent parts of the islands with a pall as of gray snow. In early stages of the eruption lakes which occupied old craters were discharged and swept down the ash-covered mountain valleys in torrents of boiling mud.

On several occasions there was shot from the crater of each volcano a thick and heavy cloud of incandescent ashes and steam, which rushed down the mountain side like an avalanche, red with glowing stones and scintillating with lightning flashes. Forests and buildings in its path were leveled as by a tornado, wood was charred and set on fire by the incandescent fragments, all vegetation was destroyed, and to breathe the steam and hot, suffocating dust of the cloud was death to every living creature. On the morning of the 8th of May, 1902, the first of these peculiar avalanches from Mt. Pelee fell on the city of St. Pierre and instantly destroyed the lives of its thirty thousand inhabitants.

The eruptions of many volcanoes partake of both the effusive and the explosive types: the molten rock in the pipe is in part blown into the air with explosions of steam, and in part is discharged in streams of lava over the lip of the crater and from fissures in the sides of the cone. Such are the eruptions of

Vesuvius, one of which is illustrated in Figure 219.

SUBMARINE ERUPTIONS. The many volcanic islands of the ocean and the coral islands resting on submerged volcanic peaks prove that eruptions have often taken place upon the ocean floor and have there built up enormous piles of volcanic fragments and lava. The Hawaiian volcanoes rise from a depth of eighteen thousand feet of water and lift their heads to about thirty thousand feet above the ocean bed. Christmas Island (see p. 194), built wholly beneath the ocean, is a coral-capped volcanic peak, whose total height, as measured from the bottom of the sea, is more than fifteen thousand feet. Deep-sea soundings have revealed the presence of numerous peaks which fail to reach sea level and which no doubt are submarine volcanoes. A number of volcanoes on the land were submarine in their early stages, as, for example, the vast pile of Etna, the celebrated Sicilian volcano, which rests on stratified volcanic fragments containing marine shells now uplifted from the sea.

Submarine outflows of lava and deposits of volcanic fragments become covered with sediments during the long intervals between eruptions. Such volcanic deposits are said to be CONTEMPORANEOUS, because they are formed during the same period as the strata among which they are imbedded. Contemporaneous lava sheets may be expected to bake the surface of the stratum on which they rest, while the sediments deposited upon them are unaltered by their heat. They are among the most permanent records of volcanic action, far outlasting the greatest volcanic mountains built in open air.

From upraised submarine volcanoes, such as Christmas Island, it is learned that lava flows which are poured out upon the bottom of the sea do not differ materially either in composition or texture from those of the land.

VOLCANIC PRODUCTS

Vast amounts of steam are, as we have seen, emitted from volcanoes, and comparatively small quantities of other vapors, such as various acid and sulphurous gases. The rocks erupted from volcanoes differ widely in chemical composition and in texture.

ACIDIC AND BASIC LAVAS. Two classes of volcanic rocks may be distinguished,--those containing a large proportion of silica (silicic acid, $SiO2$)

and therefore called ACIDIC, and those containing less silica and a larger proportion of the bases (lime, magnesia, soda, etc.) and therefore called BASIC. The acidic lavas, of which RHYOLITE and THRACHYTE are examples, are comparatively light in color and weight, and are difficult to melt. The basic lavas, of which BASALT is a type, are dark and heavy and melt at a lower temperature.

SCORIA AND PUMICE. The texture of volcanic rocks depends in part on the degree to which they were distended by the steam which permeated them when in a molten state. They harden into compact rock where the steam cannot expand. Where the steam is released from pressure, as on the surface of a lava stream, it forms bubbles (steam blebs) of various sizes, which give the hardened rock a cellular structure (Fig. 220), In this way are formed the rough slags and clinkers called SCORIA, which are found on the surface of flows and which are also thrown out as clots of lava in explosive eruptions.

On the surface of the seething lava in the throat of the volcano there gathers a rock foam, which, when hurled into the air, is cooled and falls as PUMICE,--a spongy gray rock so light that it floats on water.

AMYGDULES. The steam blebs of lava flows are often drawn out from a spherical to an elliptical form resembling that of an almond, and after the rock has cooled these cavities are gradually filled with minerals deposited from solution by underground water. From their shape such casts are called amygdules (Greek, amygdalon, an almond). Amygdules are commonly composed of silica. Lavas contain both silica and the alkalies, potash and soda, and after dissolving the alkalies, percolating water is able to take silica also into solution. Most AGATES are banded amygdules in which the silica has been laid in varicolored, concentric layers.

GLASSY AND STONY LAVAS. Volcanic rocks differ in texture according also to the rate at which they have solidified. When rapidly cooled, as on the surface of a lava flow, molten rock chills to a glass, because the minerals of which it is composed have not had time to separate themselves from the fused mixture and form crystals. Under slow cooling, as in the interior of the flow, it becomes a stony mass composed of crystals set in a glassy paste. In thin slices of volcanic glass one may see under the microscope the beginnings of crystal growth in filaments and needles and feathery forms, which are the rudiments of the crystals of various minerals.

Spherulites, which also mark the first changes of glassy lavas toward a stony condition, are little balls within the rock, varying from microscopic size to several inches in diameter, and made up of radiating fibers.

Perlitic structure, common among glassy lavas, consists of microscopic curving and interlacing cracks, due to contraction.

FLOW LINES are exhibited by volcanic rocks both to the naked eye and under the microscope. Steam blebs, together with crystals and their embryonic forms, are left arranged in lines and streaks by the currents of the flowing lava as it stiffened into rock.

PORPHYRITIC STRUCTURE. Rocks whose ground mass has scattered through it large conspicuous crystals are said to be PORPHYRITIC, and it is especially among volcanic rocks that this structure occurs. The ground mass of porphyries either may be glassy or may consist in part of a felt of minute crystals; in either case it represents the consolidation of the rock after its outpouring upon the surface. On the other hand, the large crystals of porphyry have slowly formed deep below the ground at an earlier date.

COLUMNAR STRUCTURE. Just as wet starch contracts on drying to prismatic forms, so lava often contracts on cooling to a mass of close-set, prismatic, and commonly six-sided columns, which stand at right angles to the cooling surface. The upper portion of a flow, on rapid cooling from the surface exposed to the air, may contract to a confused mass of small and irregular prisms; while the remainder forms large and beautifully regular columns, which have grown upward by slow cooling from beneath.

FRAGMENTAL MATERIALS

Rocks weighing many tons are often thrown from a volcano at the beginning of an outburst by the breaking up of the solidofied floor of the crater; and during the progress of an eruption large blocks may be torn from the throat of the volcano by the outrush of steam. But the most important fragmental materials are those derived from the lava itself. As lava rises in the pipe, the steam which permeates it is released from pressure and explodes, hurling the lava into the air in fragments of all sizes,--large pieces of scoria, LAPILLI (fragments the size of a pea or walnut), volcanic "sand" and volcanic "ashes." The latter resemble in appearance the ashes of wood or coal, but they are not in any sense, like them, a residue after combustion.

Volcanic ashes are produced in several ways: lava rising in the volcanic duct is exploded into fine dust by the steam which permeates it; glassy lava, hurled into the air and cooled suddenly, is brought into a state of high strain and tension, and, like Prince Rupert's drops, flies to pieces at the least provocation. The clash of rising and falling projectiles also produces some dust, a fair sample of which may be made by grating together two pieces of pumice.

Beds of volcanic ash occur widely among recent deposits in the western United States. In Nebraska ash beds are found in twenty counties, and are often as white as powdered pumice. The beds grow thicker and coarser toward the southwestern part of the state, where their thickness sometimes reaches fifty feet. In what direction would you look for the now extinct volcano whose explosive eruptions are thus recorded?

TUFF. This is a convenient term designating any rock composed of volcanic fragments. Coarse tuffs of angular fragments are called VOLCANIC BRECIA, and when the fragments have been rounded and sorted by water the rock is termed a VOLCANIC CONGLOMERATE. Even when deposited in the open air, as on the slopes of a volcano, tuffs may be rudely bedded and their fragments more or less rounded, and unless marine shells or the remains of land plants and animals are found as fossils in them, there is often considerable difficulty in telling whether they were laid in water or in air. In either case they soon become consolidated. Chemical deposits from percolating waters fill the interstices, and the bed of loose fragments is cemented to hard rock.

The materials of which tuffs are composed are easily recognized as volcanic in their origin. The fragments are more or less cellular, according to the degree to which they were distended with steam when in a molten state, and even in the finest dust one may see the glass or the crystals of lava from which it was derived. Tuffs often contain VOCLANIC BOMBS,--balls of lava which took shape while whirling in the air, and solidified before falling to the ground.

ANCIENT VOLCANIC ROCKS. It is in these materials and structures which we have described that volcanoes leave some of their most enduring records. Even the volcanic rocks of the earliest geological ages, uplifted after long burial beneath the sea and exposed to view by deep erosion, are recognized and their history read despite the many changes which they may have undergone. A sheet of ancient lava may be distinguished by its composition from the sediments among which it is imbedded. The direction of its flow lines may be noted. The cellular and slaggy surface where the pasty lava was distended by escaping steam is recognized by the amygdules which now fill the ancient steam blebs. In a pile of successive sheets of lava each flow may be distinguished and its thickness measured; for the surface of each sheet is glassy and scoriaceous, while beneath its upper portions the lava of each flow is more dense and stony. The length of time which elapsed before a sheet was buried beneath the materials of succeeding eruptions may be told by the amount of weathering which it had undergone, the depth of ancient soil--now baked to solid rock--upon it, and the erosion which it had suffered in the interval.

If the flow occurred from some submarine volcano, we may recognize the fact by the sea-laid sediments which cover it, filling the cracks and crevices of its upper surface and containing pieces of lava washed from it in their basal layers.

Long-buried glassy lavas devitrify, or pass to a stony condition, under the unceasing action of underground waters; but their flow lines and perlitic and spherulitic structures remain to tell of their original state.

Ancient tuffs are known by the fragmental character of their volcanic material, even though they have been altered to firm rock. Some remains of land animals and plants may be found imbedded to tell that the beds were laid in open air; while the remains of marine organisms would prove as surely that the tuffs were deposited in the sea.

In these ways ancient volcanoes have been recognized near Boston, in southeastern Pennsylvania, about Lake Superior, and in other regions of the United States.

THE LIFE HISTORY OF A VOLCANO

The invasion of a region by volcanic forces is attended by movements of the crust heralded by earthquakes. A fissure or a pipe is opened and the building of the cone or the spreading of wide lava sheets is begun.

VOLCANIC CONES. The shape of a volcanic cone depends chiefly on the materials erupted. Cones made of fragments may have sides as steep as the angle of repose, which in the case of coarse scoria is sometimes as high as thirty or forty degrees. About the base of the mountain the finer materials erupted are spread in more gentle slopes, and are also washed forward by rains and streams. The normal profile is thus a symmetric cone with a flaring base.

Cones built of lava vary in form according to the liquidity of the lava. Domes of gentle slope, as those of Hawaii, for example, are formed of basalt, which flows to long distances before it congeals. When superheated and emitted from many vents, this easily melted lava builds great plateaus, such as that of Iceland. On the other hand, lavas less fusible, or poured out at a lower temperature, stiffen when they have flowed but a short distance, and accumulate in a steep cone. Trachyte has been extruded in a state so viscid that it has formed steepsided domes like that of Sarcoui.

Most volcanoes are built, like Vesuvius, both of lava flows and of tuffs, and sections show that the structure of the cone consists of outward-dipping, alternating layers of lava, scoria, and ashes.

From time to time the cone is rent by the violence of explosions and by the weight of the column of lava in the pipe. The fissures are filled with lava and some discharge on the sides of the mountain, building parasitic cones, while all form dikes, which strengthen the pile with ribs of hard rock and make it more difficult to rend.

Great catastrophes are recorded in the shape of some volcanoes which consist of a circular rim perhaps miles in diameter, inclosing a vast crater or a caldera within which small cones may rise. We may infer that at some time the top of the mountain has been blown off, or has collapsed and been engulfed because some reservoir beneath had been emptied by long-continued eruptions.

The cone-building stage may be said to continue until eruptions of lava and fragmental materials cease altogether. Sooner or later the volcanic forces shift or die away, and no further eruptions add to the pile or replace its losses by erosion during periods of repose. Gases however are still emitted, and, as sulphur vapors are conspicuous among them, such vents are called SOLFATARAS. Mount Hood, in Oregon, is an example of a volcano sunk to this stage. From a steaming rift on its side there rise sulphurous fumes which, half a mile down the wind, will tarnish a silver coin.

GEYSERS AND HOT SPRINGS. The hot springs of volcanic regions are among the last vestiges of volcanic heat. Periodically eruptive boiling springs are termed geysers. In each of the geyser regions of the earth--the Yellowstone National Park, Iceland, and New Zealand--the ground water of the locality is supposed to be heated by ancient lavas that, because of the poor conductivity of the rock, still remain hot beneath the surface.

OLD FAITHFUL, one of the many geysers of the Yellowstone National Park, plays a fountain of boiling water a hundred feet in air; while clouds of vapor from the escaping steam ascend to several times that height. The eruptions take place at intervals of from seventy to ninety minutes. In repose the geyser is a quiet pool, occupying a craterlike depression in a conical mound some twelve feet high. The conduit of the spring is too irregular to be sounded. The mound is composed of porous silica deposited by the waters of the geyser.

Geysers erupt at intervals instead of continuously boiling, because their long, narrow, and often tortuous conduits do not permit a free circulation of the water. After an eruption the tube is refilled and the water again gradually becomes heated. Deep in the tube where it is in contact with hot lavas the water sooner or later reaches the boiling point, and bursting into steam shoots the water above it high in air.

CARBONATED SPRINGS. After all the other signs of life have gone, the ancient volcano may emit carbon dioxide as its dying breath. The springs of the region may long be charged with carbon dioxide, or carbonated, and where they rise through limestone may be expected to deposit large quantities of travertine. We should remember, however, that many carbonated springs, and many hot springs, are wholly independent of volcanoes.

THE DESTRUCTION OF THE CONE. As soon as the volcanic cone ceases to grow by eruptions the agents of erosion begin to wear it down, and the length of time that has elapsed since the period of active growth may be roughly measured by the degree to which the cone has been dissected. We infer that Mount Shasta, whose conical shape is still preserved despite the gullies one thousand feet deep which trench its sides, is younger than Mount Hood, which erosive agencies have carved to a pyramidal form. The pile of materials accumulated about a volcanic vent, no matter how vast in bulk, is at last swept entirely away. The cone of the volcano, active or extinct, is not old as the earth counts time; volcanoes are short- lived geological phenomena.

CRANDALL VOLCANO. This name is given to a dissected ancient volcano in the Yellowstone National Park, which once, it is estimated, reared its head thousands of feet above the surrounding country and greatly exceeded in bulk either Mount Shasta or Mount Etna. Not a line of the original mountain remains; all has been swept away by erosion except some four thousand feet of the base of the pile. This basal wreck now appears as a rugged region about thirty miles in diameter, trenched by deep valleys and cut into sharp peaks and precipitous ridges. In the center of the area is found the nucleus (N, Fig. 237),--a mass of coarsely crystalline rock that congealed deep in the old volcanic pipe. From it there radiate in all directions, like the spokes of a wheel, long dikes whose rock grows rapidly finer of grain as it leaves the vicinity of the once heated core. The remainder of the base of the ancient mountain is made of rudely bedded tuffs and volcanic breccia, with occasional flows of lava, some of the fragments of the breccia measuring as much as twenty feet in diameter. On the sides of canyons the breccia is carved by rain erosion to fantastic pinnacles. At different levels in the midst of these beds of tuff and lava are many old forest grounds. The stumps and trunks of the trees, now turned to stone, still in many cases stand upright where once they grew on the slopes of the mountain as it was building (Fig. 238). The great size and age of some of these trees indicate, the lapse of time between the eruption whose lavas or tuffs weathered to the soil on which they grew and the subsequent eruption which buried them beneath showers of stones and ashes.

Near the edge of the area lies Death Gulch, in which carbon dioxide is given off in such quantities that in quiet weather it accumulates in a heavy layer along the ground and suffocates the animals which may enter it.

CHAPTER XII

UNDERGROUND STRUCTURES OF IGNEOUS ORIGIN

It is because long-continued erosion lays bare the innermost anatomy of an extinct volcano, and even sweeps away the entire pile with much of the underlying strata, thus leaving the very roots of the volcano open to view, that we are able to study underground volcanic structures. With these we include, for convenience, intrusions of molten rock which have been driven upward into the crust, but which may not have succeeded in breaking way to the surface and establishing a volcano. All these structures are built of rock forced when in a fluid or pasty state into some cavity which it has found or made, and we may classify them therefore, according to the shape of the molds in which the molten rock has congealed, as (1) dikes, (2) volcanic necks, (3) intrusive sheets, and (4) intrusive masses.

DIKES. The sheet of once molten rock with which a fissure has been filled is known as a dike. Dikes are formed when volcanic cones are rent by explosions or by the weight of the lava column in the duct, and on the dissection of the pile they appear as radiating vertical ribs cutting across the layers of lava and tuff of which the cone is built. In regions undergoing deformation rocks lying deep below the ground are often broken and the fissures are filled with molten rock from beneath, which finds no outlet to the surface. Such dikes are common in areas of the most ancient rocks, which have been brought to light by long erosion.

In exceptional cases dikes may reach the length of fifty or one hundred miles. They vary in width from a fraction of a foot to even as much as three hundred feet.

Dikes are commonly more fine of grain on the sides than in the center, and may have a glassy and crackled surface where they meet the inclosing rock. Can you account for this on any principle which you have learned?

VOLCANIC NECKS. The pipe of a volcano rises from far below the base of the cone,--from the deep reservoir from which its eruptions are supplied. When the volcano has become extinct this great tube remains filled with hardened lava. It forms a cylindrical core of solid rock, except for some distance below the ancient crater, where it may contain a mass of fragments which had fallen back into the chimney after being hurled into the air.

As the mountain is worn down, this central column known as the VOLCANIC NECK is left standing as a conical hill (Fig. 240). Even when every other trace of the volcano has been swept away, erosion will not have passed below this great stalk on which the volcano was borne as a fiery flower whose site it remains to mark. In volcanic regions of deep denudation volcanic necks rise solitary and abrupt from the surrounding country as dome-shaped hills. They are marked features in the landscape in parts of Scotland and in the St. Lawrence valley about Montreal (Fig. 241).

INTRUSIVE SHEETS. Sheets of igneous rocks are sometimes found interleaved with sedimentary strata, especially in regions where the rocks have been deformed and have suffered from volcanic action. In some instances such a sheet is seen to be CONTEMPORANEOUS (p. 248). In other instances the sheet must be INTRUSIVE. The overlying stratum, as well as that beneath, has been affected by the heat of the once molten rock. We infer that the igneous rock when in a molten state was forced between the strata, much as a card may be pushed between the leaves of a closed book. The liquid wedged its way between the layers, lifting those above to make room for itself. The source of the intrusive sheet may often be traced to some dike (known therefore as the FEEDING DIKE), or to some mass of igneous rock.

Intrusive sheets may extend a score and more of miles, and, like the longest surface flows, the most extensive sheets consist of the more fusible and fluid lavas,--those of the basic class of which basalt is an example. Intrusive sheets are usually harder than the strata in which they lie and are therefore often left in relief after long denudation of the region (Fig. 315).

On the west bank of the Hudson there extends from New York Bay north for thirty miles a bold cliff several hundred feet high,-- the PALISADES OF THE HUDSON. It is the outcropping edge of a sheet of ancient igneous rock, which rests on stratified sandstones and is overlain by strata of the same series. Sandstones and lava sheet together dip gently to the west arid the latter disappears from view two miles back from the river.

It is an interesting question whether the Palisades sheet is CONTEMPORANEOUS or INTRUSIVE. Was it outpoured on the sandstones beneath it when they formed the floor of the sea, and covered forthwith by the sediments of the strata above, or was it intruded among these beds at a later date?

The latter is the case: for the overlying stratum is intensely baked along the zone of contact. At the west edge of the sheet is found the dike in which the lava rose to force its way far and wide between the strata.

ELECTRIC PEAK, one of the prominent mountains of the Yellowstone National Park, is carved out of a mass of strata into which many sheets of molten rock have been intruded. The western summit consists of such a sheet several hundred feet thick. Studying the section of Figure 244, what inference do you draw as to the source of these intrusive sheets?

INTRUSIVE MASSES

BOSSES. This name is generally applied to huge irregular masses of coarsely crystalline igneous rock lying in the midst of other formations. Bosses vary greatly in size and may reach scores of miles in extent. Seldom are there any evidences found that bosses ever had connection with the surface. On the other hand, it is often proved that they have been driven, or have melted their way, upward into the formations in which they lie; for they give off dikes and intrusive sheets, and have profoundly altered the rocks about them by their heat.

The texture of the rock of bosses proves that consolidation proceeded slowly and at great depths, and it is only because of vast denudation that they are now exposed to view. Bosses are commonly harder than the rocks about them, and stand up, therefore, as rounded hills and mountainous ridges long after the surrounding country has worn to a low plain.

The base of bosses is indefinite or undetermined, and in this respect they differ from laccoliths. Some bosses have broken and faulted the overlying beds; some have forced the rocks aside and melted them away.

The SPANISH PEAKS of southeastern Colorado were formed by the upthrust of immense masses of igneous rock, bulging and breaking the overlying strata. On one side of the mountains the throw of the fault is nearly a mile, and fragments of deep-lying beds were dragged upward by the rising masses. The adjacent rocks were altered by heat to a distance of several thousand feet. No evidence appears that the molten rock ever reached the surface, and if volcanic eruptions ever took place either in lava flows or fragmental materials, all traces of them have been effaced. The rock of the intrusive masses is coarsely crystalline, and no doubt solidified slowly under the pressure of vast thicknesses of overlying rock, now mostly removed by erosion.

A magnificent system of dikes radiates from the Peaks to a distance of fifteen miles, some now being left by long erosion as walls a hundred feet in height (Fig. 239). Intrusive sheets fed by the dikes penetrate the surrounding strata, and their edges are cut by canyons as much as twenty-five miles from the mountain. In these strata are valuable beds of lignite, an imperfect coal, which the heat of dikes and sheets has changed to coke.

LACCOLITHS. The laccolith (Greek laccos, cistern; lithos, stone) is a variety of intrusive masses in which molten rock has spread between the strata, and, lifting the strata above it to a dome-shaped form, has collected beneath them in a lens-shaped body with a flat base.

The HENRY MOUNTAINS, a small group of detached peaks in southern Utah, rise from a plateau of horizontal rocks. Some of the peaks are carved wholly in separate domelike uplifts of the strata of the plateau. In others, as Mount Hillers, the largest of the group, there is exposed on the summit a core of igneous rock from which the sedimentary rocks of the flanks dip steeply outward in all directions. In still others erosion has stripped off the covering strata and has laid bare the core to its base; and its shape is here seen to be that of a plano-convex lens or a baker's bun, its flat base resting on the undisturbed bedded rocks beneath. The structure of Mount Hillers is shown in Figure 248. The nucleus of igneous rock is four miles in diameter and more than a mile in depth.

REGIONAL INTRUSIONS. These vast bodies of igneous rock, which may reach hundreds of miles in diameter, differ little from bosses except in their immense bulk. Like bosses, regional intrusions give off dikes and sheets and greatly change the rocks about them by their heat. They are now exposed to view only because of the profound denudation which has removed the upheaved dome of rocks beneath which they slowly cooled. Such intrusions are accompanied --whether as cause or as effect is still hardly known--by deformations, and their masses of igneous rock are thus found as the core of many great mountain ranges. The granitic masses of which the Bitter Root Mountains and the Sierra Nevadas have been largely carved are each more than three hundred miles in length. Immense regional intrusions, the cores of once lofty mountain ranges, are found upon the Laurentian peneplain.

PHYSIOGRAPHIC EFFECTS OF INTRUSIVE MASSES. We have already seen examples of the topographic effects of intrusive masses in Mount Hillers, the Spanish Peaks, and in the great mountain ranges mentioned in the paragraph on regional intrusions, although in the latter instances these effects are entangled with the effects of other processes. Masses of igneous rock cannot be intruded within the crust without an accompanying deformation on a scale corresponding to the bulk of the intruded mass. The overlying strata are arched into hills or mountains, or, if the molten material is of great extent, the strata may conceivably be floated upward to the height of a plateau. We may suppose that the transference of molten matter from one region to another may be among the causes of slow subsidences and elevations. Intrusions give rise to fissures, dikes, and intrusive sheets, and these dislocations cannot fail to produce earthquakes. Where intrusive masses open communication with the surface, volcanoes are established or fissure eruptions occur such as those of Iceland.

THE INTRUSIVE ROCKS

The igneous rocks are divided into two general classes,--the VOLCANIC or ERUPTIVE rocks, which have been outpoured in open air or on the floor of the sea, and the INTRUSIVE rocks, which have been intruded within the rocks of the crust and have solidified below the surface. The two classes are alike in chemical composition and may be divided into acidic and basic groups. In texture the intrusive rocks differ from the volcanic rocks because of the different conditions under which they have solidified. They cooled far more slowly beneath the cover of the rocks into which they were pressed than is permitted to lava flows in open air. Their constituent minerals had ample opportunity to sort themselves and crystallize from the fluid mixture, and none of that mixture was left to congeal as a glassy paste.

They consolidated also under pressure. They are never scoriaceous, for the steam with which they were charged was not allowed to expand and distend them with steam blebs. In the rocks of the larger intrusive masses one may see with a powerful microscope exceedingly minute cavities, to be counted by many millions to the cubic inch, in which the gaseous water which the mass contained was held imprisoned under the immense pressure of the overlying rocks.

Naturally these characteristics are best developed in the intrusives which cooled most slowly, i.e. in the deepest-seated and largest masses; while in those which cooled more rapidly, as in dikes and sheets, we find gradations approaching the texture of surface flows.

VARIETIES OF THE INTRUSIVE ROCKS. We will now describe a few of the varieties of rocks of deep-seated intrusions. All are even grained, consisting of a mass of crystalline grains formed during one continuous stage of solidification, and no porphyritic crystals appear as in lavas.

GRANITE, as we have learned already, is composed of three minerals,--quartz, feldspar, and mica. According to the color of the feldspar the rock may be red, or pink, or gray. Hornblende--a black or dark green mineral, an iron-magnesian silicate, about as hard as feldspar--is sometimes found as a fourth constituent, and the rock is then known as HORNBLENDIC GRANITE. Granite is an acidic rock corresponding to rhyolite in chemical composition. We may believe that the same molten mass which supplies this acidic lava in surface flows solidifies as granite deep below ground in the volcanic reservoir.

SYENITE, composed of feldspar and mica, has consolidated from a less siliceous mixture than has granite.

DIORITE, still less siliceous, is composed of hornblende and feldspar,--the latter mineral being of different variety from the feldspar of granite and syenite.

GABBRO, a typical basic rock, corresponds to basalt in chemical composition. It is a dark, heavy, coarsely crystalline aggregate of feldspar and AUGITE (a dark mineral allied to hornblende). It often contains MAGNETITE (the magnetic black oxide of iron) and OLIVINE (a greenish magnesian silicate).

In the northern states all these types, and many others also of the vast number of varieties of intrusive rocks, can be found among the rocks of the drift brought from the areas of igneous rock in Canada and the states of our northern border.

SUMMARY. The records of geology prove that since the earliest of their annals tremendous forces have been active in the earth. In all the past, under pressures inconceivably great, molten rock has been driven upward into the rocks of the crust. It has squeezed into fissures forming dikes; it has burrowed among the strata as intrusive sheets; it has melted the rocks away or lifted the overlying strata, filling the chambers which it has made with intrusive masses. During all geological ages molten rock has found way to the surface, and volcanoes have darkened the sky with clouds of ashes and poured streams of glowing lava down their sides. The older strata,--the strata which have been most deeply buried,--and especially those which have suffered most from folding and from fracture, show the largest amount of igneous intrusions. The molten rock which has been driven from the earth's interior to within the crust or to the surface during geologic time must be reckoned in millions of cubic miles.

THE INTERIOR CONDITION OF THE EARTH AND CAUSES OF VULCANISM AND DEFORMATION

The problems of volcanoes and of deformation are so closely connected with that of the earth's interior that we may consider them together. Few of these problems are solved, and we may only state some known facts and the probable conclusions which may be drawn as inferences from them.

THE INTERIOR OF THE EARTH IS HOT. Volcanoes prove that in many parts of the earth there exist within reach of the surface regions of such intense heat that the rock is in a molten condition. Deep wells and mines show everywhere an increase in temperature below the surface shell affected by the heat of summer and the cold of winter,--a shell in temperate latitudes sixty or seventy feet thick. Thus in a boring more than a mile deep at Schladebach, Germany, the earth grows warmer at the rate of 1 degrees F. for every sixty-seven feet as we descend. Taking the average rate of increase at one degree for every sixty feet of descent, and assuming that this rate, observed at the moderate distances open to observation, continues to at least thirty-five miles, the temperature at that depth must be more than three thousand degrees,--a temperature at which all ordinary rocks would melt at the earth's surface. The rate of increase in temperature probably lessens as we go downward, and it may not be appreciable below a few hundred miles. But there is no reason to doubt that THE INTERIOR OF THE EARTH IS INTENSELY HOT. Below a depth of one or two score miles we may imagine the rocks everywhere glowing with heat.

Although the heat of the interior is great enough to melt all rocks at atmospheric pressure, it does not follow that the interior is fluid. Pressure raises the fusing point of rocks, and the weight of the crust may keep the interior in what may be called a solid state, although so hot as to be a liquid or a gas were the pressure to be removed.

THE INTERIOR OF THE EARTH IS RIGID AND HEAVY. The earth behaves as a globe more rigid than glass under the attractions of the sun and moon. It is not deformed by these stresses as is the ocean in the tides, proving that it is not a fluid ball covered with a yielding crust a few miles thick. Earthquakes pass through the earth faster than they would were it of solid steel. Hence the rocks of the interior are highly elastic, being brought by pressure to a compact, continuous condition unbroken by the cracks and vesicles of surface rocks. THE INTERIOR OF THE EARTH IS RIGID

The common rocks of the crust are about two and a half times heavier than water, while the earth as a whole weighs five and six-tenths times as much as a globe of water of the same size. THE INTERIOR IS THEREFORE MUCH MORE HEAVY THAN THE CRUST. This may be caused in part by compression of the interior under the enormous weight of the crust, and in part also by an assortment of material, the heavier substances, such as the heavy metals, having gravitated towards the center.

Between the crust, which is solid because it is cool, and the interior, which is hot enough to melt were it not for the pressure which keeps it dense and rigid, there may be an intermediate zone in which heat and pressure are so evenly balanced that here rock liquefies whenever and wherever the pressure upon it may be relieved by movements of the crust. It is perhaps from such a subcrustal layer that the lava of volcanoes is supplied.

THE CAUSES OF VOLCANIC ACTION. It is now generally believed that the HEAT of volcanoes is that of the earth's interior. Other causes, such as friction and crushing in the making of mountains and the chemical reactions between oxidizing agents of the crust and the unoxidized interior, have been suggested, but to most geologists they seem inadequate.

There is much difference of opinion as to the FORCE which causes molten rock to rise to the surface in the ducts of volcanoes. Steam is so evidently concerned in explosive eruptions that many believe that lava is driven upward by the expansive force of the steam with which it is charged, much as a viscid liquid rises and boils over in a test tube or kettle.

But in quiet eruptions, and still more in the irruption of intrusive sheets and masses, there is little if any evidence that steam is the driving force. It is therefore believed by many geologists that it is PRESSURE DUE TO CRUSTAL MOVEMENTS AND INTERNAL STRESSES which squeezes molten rock from below into fissures and ducts in the crust. It is held by some that where considerable water is supplied to the rising column of lava, as from the ground water of the surrounding region, and where the lava is viscid so that steam does not readily escape, the eruption is of the explosive type; when these conditions do not obtain, the lava outwells quietly, as in the Hawaiian volcanoes. It is held by others not only that volcanoes are due to the outflow of the earth's deep-seated heat, but also that the steam and other emitted gases are for the most part native to the earth's interior and never have had place in the circulation of atmospheric and ground waters.

VOLCANIC ACTION AND DEFORMATION. Volcanoes do not occur on wide plains or among ancient mountains. On the other hand, where movements of the earth's crust are in progress in the uplift of high plateaus, and still more in mountain making, molten rock may reach the surface, or may be driven upward toward it forming great intrusive masses. Thus extensive lava flows accompanied the upheaval of the block mountains of western North America and the uplift of the Colorado plateau. A line of recent volcanoes may be traced along the system of rift valleys which extends from the Jordan and Dead Sea through eastern Africa to Lake Nyassa. The volcanoes of the Andes show how conspicuous volcanic action may be in young rising ranges. Folded mountains often show a core of igneous rock, which by long erosion has come to form the axis and the highest peaks of the range, as if the molten rock had been squeezed up under the rising upfolds. As we decipher the records of the rocks in historical geology we shall see more fully how, in all the past, volcanic action has characterized the periods of great crustal movements, and how it has been absent when and where the earth's crust has remained comparatively at rest.

THE CAUSES OF DEFORMATION. As the earth's interior, or nucleus, is highly heated it must be constantly though slowly losing its heat by conduction through the crust and into space; and since the nucleus is cooling it must also be contracting. The nucleus has contracted also because of the extrusion of molten matter, the loss of constituent gases given off in volcanic eruptions, and (still more important) the compression and consolidation of its material under gravity. As the nucleus contracts, it tends to draw away from the cooled and solid crust, and the latter settles, adapting itself to the shrinking nucleus much as the skin of a withering apple wrinkles down upon the shrunken fruit. The unsupported weight of the spherical crust develops enormous tangential pressures, similar to the stresses of an arch or dome, and when these lateral thrusts accumulate beyond the power of resistance the solid rock is warped and folded and broken.

Since the planet attained its present mass it has thus been lessening in volume. Notwithstanding local and relative upheavals the earth's surface on the whole has drawn nearer and nearer to the center. The portions of the lithosphere which have been carried down the farthest have received the waters of the oceans, while those portions which have been carried down the least have emerged as continents.

Although it serves our convenience to refer the movements of the crust to the sea level as datum plane, it is understood that this level is by no means fixed. Changes in the ocean basins increase or reduce their capacity and thus lower or raise the level of the sea. But since these basins are connected, the effect of any change upon the water level is so distributed that it is far less noticeable than a corresponding change would be upon the land.

CHAPTER XIII

METAMORPHISM AND MINERAL VEINS

Under the action of internal agencies rocks of all kinds may be rendered harder, more firmly cemented, and more crystalline. These processes are known as METAMORPHISM, and the rocks affected, whether originally sedimentary or igneous, are called METAMORPHIC ROCKS. We may contrast with metamorphism the action of external agencies in weathering, which render rocks less coherent by dissolving their soluble parts and breaking down their crystalline grains.

CONTACT METAMORPHISM. Rocks beneath a lava flow or in contact with igneous intrusions are found to be metamorphosed to various degrees by the heat of the cooling mass. The adjacent strata may be changed only in color, hardness, and texture. Thus, next to a dike, bituminous coal may be baked to coke or anthracite, and chalk and limestone to crystalline marble. Sandstone may be converted into quartzite, and shale into ARGILLITE, a compact, massive clay rock. New minerals may also be developed. In sedimentary rocks there may be produced crystals of mica and of GARNET (a mineral as hard as quartz, commonly occurring in red, twelve-sided crystals). Where the changes are most profound, rocks may be wholly made over in structure and mineral composition.

In contact metamorphism, thin sheets of molten rock produce less effect than thicker ones. The strongest heat effects are naturally caused by bosses and regional intrusions, and the zone of change about them may be several miles in width. In these changes heated waters and vapors from the masses of igneous rocks undoubtedly play a very important part.

Which will be more strongly altered, the rocks about a closed dike in which lava began to cool as soon as it filled the fissure, or the rocks about a dike which opened on the surface and through which the molten rock flowed for some time?

Taking into consideration the part played by heated waters, which will produce the most far-reaching metamorphism, dikes which cut across the bedding planes or intrusive sheets which are thrust between the strata?

REGIONAL METAMORPHISM. Metamorphic rocks occur wide-spread in many regions, often hundreds of square miles in area, where such extensive changes cannot be accounted for by igneous intrusions. Such are the dissected cores of lofty mountains, as the Alps, and the worn-down bases of ancient ranges, as in New England, large areas in the Piedmont Belt, and the Laurentian peneplain.

In these regions the rocks have yielded to immense pressure. They have been folded, crumpled, and mashed, and even their minute grains, as one may see with a microscope, have often been puckered, broken, and crushed to powder. It is to these mechanical movements and strains which the rocks have suffered in every part that we may attribute their metamorphism, and the degree to which they have been changed is in direct proportion to the degree to which they have been deformed and mashed.

Other factors, however, have played important parts. Rock crushing develops heat, and allows a freer circulation of heated waters and vapors. Thus chemical reactions are greatly quickened; minerals are dissolved and redeposited in new positions, or their chemical constituents may recombine in new minerals, entirely changing the nature of the rock, as when, for example, feldspar recrystallizes as quartz and mica.

Early stages of metamorphism are seen in SLATE. Pressure has hardened the marine muds, the arkose, or the volcanic ash from which slates are derived, and has caused them to cleave by the rearrangement of their particles.

Under somewhat greater pressure, slate becomes PHYLLITE, a clay slate whose cleavage surfaces are lustrous with flat-lying mica flakes. The same pressure which has caused the rock to cleave has set free some of its mineral constituents along the cleavage planes to crystallize there as mica.

FOLIATION. Under still stronger pressure the whole structure of the rock is altered. The minerals of which it is composed, and the new minerals which develop by heat and pressure, arrange themselves along planes of cleavage or of shear in rudely parallel leaves, or FOLIA. Of this structure, called FOLIATION, we may distinguish two types,--a coarser feldspathic type, and a fine type in which other minerals than feldspar predominate.

GNEISS is the general name under which are comprised coarsely foliated rocks banded with irregular layers of feldspar and other minerals. The gneisses appear to be due in many cases to the crushing and shearing of deep-seated igneous rocks, such as granite and gabbro.

THE CRYSTALLINE SCHISTS, representing the finer types of foliation, consist of thin, parallel, crystalline leaves, which are often remarkably crumpled. These folia can be distinguished from the laminae of sedimentary rocks by their lenticular form and lack of continuity, and especially by the fact that they consist of platy, crystalline grains, and not of particles rounded by wear.

MICA SCHIST, the most common of schists, and in fact of all metamorphic rocks, is composed of mica and quartz in alternating wavy folia. All gradations between it and phyllite may be traced, and in many cases we may prove it due to the metamorphism of slates and shales. It is widespread in New England and along the eastern side of the Appalachians. TALC SCHIST consists of quartz and TALC, a light-colored magnesian mineral of greasy feel, and so soft that it can be scratched with the thumb nail.

HORNBLENDE SCHIST, resulting in many cases from the foliation of basic igneous rocks, is made of folia of hornblende alternating with bands of quartz and feldspar. Hornblende schist is common over large areas in the Lake Superior region.

QUARTZ SCHIST is produced from quartzite by the development of fine folia of mica along planes of shear. All gradations may be found between it and unfoliated quartzite on the one hand and mica schist on the other.

Under the resistless pressure of crustal movements almost any rocks, sandstones, shales, lavas of all kinds, granites, diorites, and gabbros may be metamorphosed into schists by crushing and shearing. Limestones, however, are metamorphosed by pressure into marble, the grains of carbonate of lime recrystallizing freely to interlocking crystals of calcite.

These few examples must suffice of the great class of metamorphic rocks. As we have seen, they owe their origin to the alteration of both of the other classes of rocks--the sedimentary and the igneous--by heat and pressure, assisted usually by the presence of water. The fact of change is seen in their hardness arid cementation, their more or less complete recrystallization, and their foliation; but the change is often so complete that no trace of their original structure and mineral composition remains to tell whether the rocks from which they were derived were sedimentary or igneous, or to what variety of either of these classes they belonged.

In many cases, however, the early history of a metamorphic rock can be deciphered. Fossils not wholly obliterated may prove it originally water-laid. Schists may contain rolled-out pebbles, showing their derivation from a conglomerate. Dikes of igneous rocks may be followed into a region where they have been foliated by pressure. The most thoroughly metamorphosed rocks may sometimes be traced out into unaltered sedimentary or igneous rocks, or among them may be found patches of little change where their history maybe read.

Metamorphism is most common among rocks of the earlier geological ages, and most rare among rocks of recent formation. No doubt it is now in progress where deep-buried sediments are invaded by heat either from intrusive igneous masses or from the earth's interior, or are suffering slow deformation under the thrust of mountain-making forces.

Suggest how rocks now in process of metamorphism may sometimes be exposed to view. Why do metamorphic rocks appear on the surface to-day?

MINERAL VEINS

In regions of folded and broken rocks fissures are frequently found to be filled with sheets of crystalline minerals deposited from solution by underground water, and fissures thus filled are known as mineral veins. Much of the importance of mineral veins is due to the fact that they are often metalliferous, carrying valuable native metals and metallic ores disseminated in fine particles, in strings, and sometimes in large masses in the midst of the valueless nonmetallic minerals which make up what is known as the VEIN STONE.

The most common vein stones are QUARTZ and CALCITE. FLUORITE (calcium fluoride), a mineral harder than calcite and crystallizing in cubes of various colors, and BARITE (barium sulphate), a heavy white mineral, are abundant in many veins.

The gold-bearing quartz veins of California traverse the metamorphic slates of the Sierra Nevada Mountains. Below the zone of solution (p. 45) these veins consist of a vein stone of quartz mingled with pyrite (p. 13), the latter containing threads and grains of native gold. But to the depth of about fifty feet from the surface the pyrite of the vein has been dissolved, leaving a rusty, cellular quartz with grains of the insoluble gold scattered through it.

The PLACER DEPOSITS of California and other regions are gold- bearing deposits of gravel and sand in river beds. The heavy gold is apt to be found mostly near or upon the solid rock, and its grains, like those of the sand, are always rounded. How the gold came in the placers we may leave the pupil to suggest.

Copper is found in a number of ores, and also in the native metal. Below the zone of surface changes the ore of a copper vein is often a double sulphide of iron and copper called CHALCOPYRITE, a mineral softer than pyrite--it can easily be scratched with a knife--and deeper yellow in color. For several score of feet below the ground the vein may consist of rusty quartz from which the metallic ores have been dissolved; but at the base of the zone of solution we may find exceedingly rich deposits of copper ores,-- copper sulphides, red and black copper oxides, and green and blue copper carbonates, which have clearly been brought down in solution from the leached upper portion of the vein.

ORIGIN OF MINERAL VEINS. Both vein stones and ores have been deposited slowly from solution in water, much as crystals of salt are deposited on the sides of a jar of saturated brine. In our study of underground water we learned that it is everywhere circulating through the permeable rocks of the crust, descending to profound depths under the action of gravity and again driven to the surface by hydrostatic pressure. Now fissures, wherever they occur, form the trunk channels of the underground circulation. Water descends from the surface along these rifts; it moves laterally from either side to the fissure plane, just as ground water seeps through the surrounding rocks from every direction to a well; and it ascends through these natural water ways as in an artesian well, whenever they intersect an aquifer in which water is under hydrostatic pressure.

The waters which deposit vein stones and ores are commonly hot, and in many cases they have derived their heat from intrusions of igneous rock still uncooled within the crust. The solvent power of the water is thus greatly increased, and it takes up into solution various substances from the igneous and sedimentary rocks which it traverses. For various reasons these substances stances are deposited in the vein as ores and vein stones. On rising through the fissure the water cools and loses pressure, and its capacity to hold minerals in solution is therefore lessened. Besides, as different currents meet in the fissure, some ascending, some descending, and some coming in from the sides, the chemical reaction of these various weak solutions upon one another and upon the walls of the vein precipitates the minerals of vein stuffs and ores.

As an illustration of the method of vein deposits we may cite the case of a wooden box pipe used in the Comstock mines, Nevada, to carry the hot water of the mine from one level to another, which in ten years was lined with calcium carbonate more than half an inch thick.

The Steamboat Springs, Nevada, furnish examples of mineral veins in process of formation. The steaming water rises through fissures in volcanic rocks and is now depositing in the rifts a vein stone of quartz, with metallic ores of iron, mercury, lead, and other metals.

RECONCENTRATION. Near the base of the zone of solution veins are often stored with exceptionally large and valuable ore deposits. This local enrichment of the vein is due to the reconcentration of its metalliferous ores. As the surface of the land is slowly lowered by weathering and running water, the zone of solution is lowered at an equal rate and encroaches constantly on the zone of cementation. The minerals of veins are therefore constantly being dissolved along their upper portions and carried down the fissures by ground water to lower levels, where they are redeposited.

Many of the richest ore deposits are thus due to successive concentrations: the ores were leached originally from the rocks to a large extent by laterally seeping waters; they were concentrated in the ore deposits of the vein chiefly by ascending currents; they have been reconcentrated by descending waters in the way just mentioned.

THE ORIGINAL SOURCE OF THE METALS. It is to the igneous rocks that we may look for the original source of the metals of veins. Lavas contain minute percentages of various metallic compounds, and no doubt this was the case also with the igneous rocks which formed the original earth crust. By the erosion of the igneous rocks the metals have been distributed among sedimentary strata, and even the sea has taken into solution an appreciable amount of gold and other metals, but in this widely diffused condition they are wholly useless to man. The concentration which has made them available is due to the interaction of many agencies. Earth movements fracturing deeply the rocks of the crust, the intrusion of heated masses, the circulation of underground waters, have all cooperated in the concentration of the metals of mineral veins.

While fissure veins are the most important of mineral veins, the latter term is applied also to any water way which has been filled by similar deposits from solution. Thus in soluble rocks, such as limestones, joints enlarged by percolating water are sometimes filled with metalliferous deposits, as, for example, the lead and zinc deposits of the upper Mississippi valley. Even a porous aquifer may be made the seat of mineral deposits, as in the case of some copper-bearing and silver-bearing sandstones of New Mexico.

PART III

HISTORICAL GEOLOGY

CHAPTER XIV

THE GEOLOGICAL RECORD

WHAT A FORMATION RECORDS. We have already learned that each individual body of stratified rock, or formation, constitutes a record of the time when it was laid. The structure and the character of the sediments of each formation tell whether the area was land or sea at the time when they were spread; and if the former, whether the land was river plain, or lake bed, or was covered with wind-blown sands, or by the deposits of an ice sheet. If the sediments are marine, we may know also whether they were laid in shoal water near the shore or in deeper water out at sea, and whether during a period of emergence, or during a period of subsidence when the sea transgressed the land. By the same means each formation records the stage in the cycle of erosion of the land mass from which its sediments were derived. An unconformity between two marine formations records the fact that between the periods when they were deposited in the sea the area emerged as land and suffered erosion. The attitude and structure of the strata tell also of the foldings and fractures, the deformation and the metamorphism, which they have suffered; and the igneous rocks associated with them as lava flows and igneous intrusions add other details to the story. Each formation is thus a separate local chapter in the geological history of the earth, and its strata are its leaves. It contains an authentic record of the physical conditions--the geography--of the time and place when and where its sediments were laid.

PAST CYCLES OF EROSION. These chapters in the history of the planet are very numerous, although much of the record has been destroyed in various ways. A succession of different formations is usually seen in any considerable section of the crust, such as a deep canyon or where the edges of upturned strata are exposed to view on the flanks of mountain ranges; and in any extensive area, such as a state of the Union or a province of Canada, the number of formations outcropping on the surface is large.

It is thus learned that our present continent is made up for. the most part of old continental deltas. Some, recently emerged as the strata of young coastal plains, are the records of recent cycles of erosion; while others were deposited in the early history of the earth, and in many instances have been crumpled into mountains, which afterwards were leveled to their bases and lowered beneath the sea to receive a cover of later sediments before they were again uplifted to form land.

The cycle of erosion now in progress and recorded in the layers of stratified rock being spread beneath the sea in continental deltas has therefore been preceded by many similar cycles. Again and again movements of the crust have brought to an end one cycle-- sometimes when only well under way, and sometimes when drawing toward its close--and have begun another. Again and again they have added to the land areas which before were sea, with all their deposition records of earlier cycles, or have lowered areas of land beneath the sea to receive new sediments.

THE AGE OF THE EARTH. The thickness of the stratified rocks now exposed upon the eroded surface of the continents is very great. In the Appalachian region the strata are seven or eight miles thick, and still greater thicknesses have been measured in several other mountain ranges. The aggregate thickness of all the formations of the stratified rocks of the earth's crust, giving to each formation its maximum thickness wherever found, amounts to not less than forty miles. Knowing how slowly sediments accumulate upon the sea floor, we must believe that the successive cycles which the earth has seen stretch back into a past almost inconceivably remote, and measure tens of millions and perhaps even hundreds of millions of years.

HOW THE FORMATIONS ARE CORRELATED AND THE GEOLOGICAL RECORD MADE UP. Arranged in the order of their succession, the formations of the earth's crust would constitute a connected record in which the geological history of the planet may be read, and therefore known as the GEOLOGICAL RECORD. But to arrange the formations in their natural order is not an easy task. A complete set of the volumes of the record is to be found in no single region. Their leaves and chapters are scattered over the land surface of the globe. In one area certain chapters may be found, though perhaps with many missing leaves, and with intervening chapters wanting, and these absent parts perhaps can be supplied only after long search through many other regions.

Adjacent strata in any region are arranged according to the LAW OF SUPERPOSITION, i.e. any stratum is younger than that on which it was deposited, just as in a pile of paper, any sheet was laid later than that on which it rests. Where rocks have been disturbed, their original attitude must be determined before the law can be applied. Nor can the law of superposition be used in identifying and comparing the strata of different regions where the formations cannot be traced continuously from one region to the other.

The formations of different regions are arranged in their true order by the LAW OF INCLUDED ORGANISMS; i.e. formations, however widely separated, which contain a similar assemblage of fossils are equivalent and belong to the same division of geological time.

The correlation of formations by means of fossils may be explained by the formations now being deposited about the north Atlantic. Lithologically they are extremely various. On the continental shelf of North America limestones of different kinds are forming off Florida, and sandstones and shales from Georgia northward. Separated from them by the deep Atlantic oozes are other sedimentary deposits now accumulating along the west coast of Europe. If now all these offshore formations were raised to open air, how could they be correlated? Surely not by lithological likeness, for in this respect they would be quite diverse. All would be similar, however, in the fossils which they contain. Some fossil species would be identical in all these formations and others would be closely allied. Making all due allowance for differences in species due to local differences in climate and other physical causes, it would still be plain that plants and animals so similar lived at the same period of time, and that the formations in which their remains were imbedded were contemporaneous in a broad way. The presence of the bones of whales and other marine mammals would prove that the strata were laid after the appearance of mammals upon earth, and imbedded relics of man would give a still closer approximation to their age. In the same way we correlate the earlier geological formations.

For example, in 1902 there were collected the first fossils ever found on the antarctic continent. Among the dozen specimens obtained were some fossil ammonites (a family of chambered shells) of genera which are found on other continents in certain formations classified as the Cretaceous system, and which occur neither above these formations nor below them. On the basis of these few fossils we may be confident that the strata in which they were found in the antarctic region were laid in the same period of geologic time as were the Cretaceous rocks of the United States and Canada.

THE RECORD AS A TIME SCALE. By means of the law of included organisms and the law of superposition the formations of different countries and continents are correlated and arranged in their natural order. When the geological record is thus obtained it may be used as a universal time scale for geological history. Geological time is separated into divisions corresponding to the times during which the successive formations were laid. The largest assemblages of formations are known as groups, while the corresponding divisions of time are known as eras. Groups are subdivided into systems, and systems into series. Series are divided into stages and substages,--subdivisions which do not concern us in this brief treatise. The corresponding divisions of time are given in the following table.

STRATA TIME Group Era System Period Series Epoch

The geologist is now prepared to read the physical history--the geographical development--of any country or of any continent by means of its formations, when he has given each formation its true place in the geological record as a time scale.

The following chart exhibits the main divisions of the record, the name given to each being given also to the corresponding time division. Thus we speak of the CAMBRIAN SYSTEM, meaning a certain succession of formations which are classified together because of broad resemblances in their included organisms; and of the CAMBRIAN PERIOD, meaning the time during which these rocks were deposited.

Group and Era System and Period Series and Epoch

|Quaternary-----|Recent Cenozoic------| |Pleistocene | |Tertiary--------|Pliocene |Miocene |Eocene |Cretaceous Mesozoic------|Jurassic |Triassic

|Permian |Carboniferous--|Pennsylvanian | |Mississippian Paleozoic----- |Devonian |Silurian |Ordovician |Cambrian

Algonkian Archean

FOSSILS AND WHAT THEY TEACH

The geological formations contain a record still more important than that of the geographical development of the continents; the fossils imbedded in the rocks of each formation tell of the kinds of animals and plants which inhabited the earth at that time, and from these fossils we are therefore able to construct the history of life upon the earth.

FOSSILS. These remains of organisms are found in the strata in all degrees of perfection, from trails and tracks and fragmentary impressions, to perfectly preserved shells, wood, bones, and complete skeletons. As a rule, it is only the hard parts of animals and plants which have left any traces in the rocks. Sometimes the original hard substance is preserved, but more often it has been replaced by some less soluble material. Petrifaction, as this process of slow replacement is called, is often carried on in the most exquisite detail. When wood, for example, is undergoing petrifaction, the woody tissue may be replaced, particle by particle, by silica in solution through the action of underground waters, even the microscopic structures of the wood being perfectly reproduced. In shells originally made of ARAGONITE, a crystalline form of carbonate of lime, that mineral is usually replaced by CALCITE, a more stable form of the same substance. The most common petrifying materials are calcite, silica, and pyrite.

Often the organic substance has neither been preserved nor replaced, but the FORM has been retained by means of molds and casts. Permanent impressions, or molds, may be made in sediments not only by the hard parts of organisms, but also by such soft and perishable parts as the leaves of plants, and, in the rarest instances, by the skin of animals and the feathers of birds. In fine-grained limestones even the imprints of jellyfish have been retained.

The different kinds of molds and casts may be illustrated by means of a clam shell and some moist clay, the latter representing the sediments in which the remains of animals and plants are entombed. Imbedding the shell in the clay and allowing the clay to harden, we have a MOLD OF THE EXTERIOR of the shell, as is seen on cutting the clay matrix in two and removing the shell from it. Filling this mold with clay of different color, we obtain a CAST OF THE EXTERIOR, which represents accurately the original form and surface markings of the shell. In nature, shells and other relics of animals or plants are often removed by being dissolved by percolating waters, and the molds are either filled with sediments or with minerals deposited from solution.

Where the fossil is hollow, a CAST OF THE INTERIOR is made in the same way. Interior casts of shells reproduce any markings on the inside of the valves, and casts of the interior of the skulls of ancient vertebrates show the form and size of their brains.

IMPERFECTION OF THE LIFE RECORD. At the present time only the smallest fraction of the life on earth ever gets entombed in rocks now forming. In the forest great fallen tree trunks, as well as dead leaves, decay, and only add a little to the layer of dark vegetable mold from which they grew. The bones of land animals are, for the most part, left unburied on the surface and are soon destroyed by chemical agencies. Even where, as in the swamps of river, flood plains and in other bogs, there are preserved the remains of plants, and sometimes insects, together with the bones of some animal drowned or mired, in most cases these swamp and bog deposits are sooner or later destroyed by the shifting channels of the stream or by the general erosion of the land.

In the sea the conditions for preservation are more favorable than on land; yet even here the proportion of animals and plants whose hard parts are fossilized is very small compared with those which either totally decay before they are buried in slowly accumulating sediments or are ground to powder by waves and currents.

We may infer that during each period of the past, as at the present, only a very insignificant fraction of the innumerable organisms of sea and land escaped destruction and left in continental and oceanic deposits permanent records of their existence. Scanty as these original life records must have been, they have been largely destroyed by metamorphism of the rocks in which they were imbedded, by solution in underground waters, and by the vast denudation under which the sediments of earlier periods have been eroded to furnish materials for the sedimentary records of later times. Moreover, very much of what has escaped destruction still remains undiscovered. The immense bulk of the stratified rocks is buried and inaccessible, and the records of the past which it contains can never be known. Comparatively few outcrops have been thoroughly searched for fossils. Although new species are constantly being discovered, each discovery may be considered as the outcome of a series of happy accidents,--that the remains of individuals of this particular species happened to be imbedded and fossilized, that they happened to escape destruction during long ages, and that they happened to be exposed and found.

SOME INFERENCES FROM THE RECORDS OF THE HISTORY OF LIFE UPON THE PLANET. Meager as are these records, they set forth plainly some important truths which we will now briefly mention.

1. Each series of the stratified rocks, except the very deepest, contains vestiges of life. Hence THE EARTH WAS TENANTED BY LIVING CREATURES FOR AN UNCALCULATED LENGTH OF TIME BEFORE HUMAN HISTORY BEGAN.

2. LIFE ON THE EARTH HAS BEEN EVERCHANGING. The youngest strata hold the remains of existing species of animals and plants and those of species and varieties closely allied to them. Strata somewhat older contain

fewer existing species, and in strata of a still earlier, but by no means an ancient epoch, no existing species are to be found; the species of that epoch and of previous epochs have vanished from the living world. During all geological time since life began on earth old species have constantly become extinct and with them the genera and families to which they belong, and other species, genera, and families have replaced them. The fossils of each formation differ on the whole from those of every other. The assemblage of animals and plants (the FAUNA- FLORA) of each epoch differs from that of every other epoch.

In many cases the extinction of a type has been gradual; in other instances apparently abrupt. There is no evidence that any organism once become extinct has ever reappeared. The duration of a species in time, or its "vertical range" through the strata, varies greatly. Some species are limited to a stratum a few feet in thickness; some may range through an entire formation and be found but little modified in still higher beds. A formation may thus often be divided into zones, each characterized by its own peculiar species. As a rule, the simpler organisms have a longer duration as species, though not as individuals, than the more complex.

3. THE LARGER ZOOLOGICAL AND BOTANICAL GROUPINGS SURVIVE LONGER THAN THE SMALLER. Species are so short-lived that a single geological epoch may be marked by several more or less complete extinctions of the species of its fauna-flora and their replacement by other species. A genus continues with new species after all the species with which it began have become extinct. Families survive genera, and orders families. Classes are so long- lived that most of those which are known from the earliest formations are represented by living forms, and no sub-kingdom has ever become extinct.

Thus, to take an example from the stony corals,--the ZOANTHARIA,-- the particular characters--which constituted a certain SPECIES-- Facosites niagarensis--of the order are confined to the Niagara series. Its GENERIC characters appeared in other species earlier in the Silurian and continued through the Devonian. Its FAMILY characters, represented in different genera and species, range from the Ordovician to the close of the Paleozoic; while the characters which it shares with all its order, the Zoantharia, began in the Cambrian and are found in living species.

4. THE CHANGE IN ORGANISMS HAS BEEN GRADUAL. The fossils of each life zone and of each formation of a conformable series closely resemble, with some explainable exceptions, those of the beds immediately above and below. The animals and plants which tenanted the earth during any geological epoch are so closely related to those of the preceding and the succeeding epochs that we may consider them to be the descendants of the one and the ancestors of the other, thus accounting for the resemblance by heredity. It is therefore believed that the species of animals and plants now living on the earth are the descendants of the species whose remains we find entombed in the rocks, and that the chain of life has been unbroken since its beginning.

5. THE CHANGE IN SPECIES HAS BEEN A GRADUAL DIFFERENTIATION. Tracing the lines of descent of various animals and plants of the present backward through the divisions of geologic time, we find that these lines of descent converge and unite in simpler and still simpler types. The development of life may be represented by a tree whose trunk is found in the earliest ages and whose branches spread and subdivide to the growing twigs of present species.

6. THE CHANGE IN ORGANISMS THROUGHOUT GEOLOGIC TIME HAS BEEN A PROGRESSIVE CHANGE. In the earliest ages the only animals and plants on the earth were lowly forms, simple and generalized in structure; while succeeding ages have been characterized by the introduction of types more and more specialized and complex, and therefore of higher rank in the scale of being. Thus the Algonkian contains the remains of only the humblest forms of the invertebrates. In the Cambrian, Ordovician, and Silurian the invertebrates were represented in all their subkingdoms by a varied fauna. In the Devonian, fishes--the lowest of the vertebrates--became abundant. Amphibians made their entry on the stage in the Carboniferous, and reptiles came to rule the world in the Mesozoic. Mammals culminated in the Tertiary in strange forms which became more and more like those of the present as the long ages of that era rolled on; and latest of all appeared the noblest product of the creative process, man.

Just as growth is characteristic of the individual life, so gradual, progressive change, or evolution, has characterized the history of life upon the planet. The evolution of the organic kingdom from its primitive germinal forms to the complex and highly organized fauna-flora of to-day may be compared to the growth of some noble oak as it rises from the acorn, spreading loftier and more widely extended branches as it grows.

7. While higher and still higher types have continually been evolved, until man, the highest of all, appeared, THE LOWER AND EARLIER TYPES HAVE GENERALLY PERSISTED. Some which reached their culmination early in the history of the earth have since changed only in slight adjustments to a changing environment. Thus the brachiopods, a type of shellfish, have made no progress since the Paleozoic, and some of their earliest known genera are represented by living forms hardly to be distinguished from their ancient ancestors. The lowest and earliest branches of the tree of life have risen to no higher levels since they reached their climax of development long ago.

8. A strange parallel has been found to exist between the evolution of organisms and the development of the individual. In the embryonic stages of its growth the individual passes swiftly through the successive stages through which its ancestors evolved during the millions of years of geologic time. THE DEVELOPMENT OF THE INDIVIDUAL RECAPITULATES THE EVOLUTION OF THE RACE.

The frog is a typical amphibian. As a tadpole it passes through a stage identical in several well-known features with the maturity of fishes; as, for example, its aquatic life, the tail by which it swims, and the gills through which it breathes. It is a fair inference that the tadpole stage in the life history of the frog represents a stage in the evolution of its kind,--that the Amphibia are derived from fishlike ancestral forms. This inference is amply confirmed in the geological record; fishes appeared before Amphibia and were connected with them by transitional forms.

THE GREAT LENGTH OF GEOLOGIC TIME INFERRED FROM THE SLOW CHANGE OF SPECIES. Life forms, like land forms, are thus subject to change under the influence of their changing environment and of forces acting from within. How slowly they change may be seen in the apparent stability of existing species. In the lifetime of the observer and even in the recorded history of man, species seem as stable as the mountain and the river. But life forms and land forms are alike variable, both in nature and still more under the shaping hand of man. As man has modified the face of the earth with his great engineering works, so he has produced widely different varieties of many kinds of domesticated plants and animals, such as the varieties of the dog and the horse, the apple and the rose, which may be regarded in some respects as new species in the making. We have assumed that land forms have changed in the past under the influence of forces now in operation. Assuming also that life forms have always changed as they are changing at present, we come to realize something of the immensity of geologic time required for the evolution of life from its earliest lowly forms up to man.

It is because the onward march of life has taken the same general course the world over that we are able to use it as a UNIVERSAL TIME SCALE and divide geologic time into ages and minor subdivisions according to the ruling or characteristic organisms then living on the earth. Thus, since vertebrates appeared, we have in succession the Age of Fishes, the Age of Amphibians, the Age of Reptiles, and the Age of Mammals.

The chart given on page 295 is thus based on the law of superposition and the law of the evolution of organisms. The first law gives the succession of the formations in local areas. The fossils which they contain demonstrate the law of the progressive appearance of organisms, and by means of this law the formations of different countries are correlated and set each in its place in a universal time scale and grouped together according to the affinities of their imbedded organic remains.

GEOLOGIC TIME DIVISIONS COMPARED WITH THOSE OF HUMAN HISTORY. We may compare the division of geologic time into eras, periods, and other divisions according to the dominant life of the time, to the ill-defined ages into which human history is divided according to the dominance of some nation, ruler, or other characteristic feature. Thus we speak of the DARK AGES, the AGE OF ELIZABETH, and the AGE OF ELECTRICITY. These crude divisions would be of much value if, as in the case of geologic time, we had no exact reckoning of human history by years.

And as the course of human history has flowed in an unbroken stream along quiet reaches of slow change and through periods of rapid change and revolution, so with the course of geologic history. Periods of quiescence, in which revolutionary forces are perhaps gathering head, alternate with periods of comparatively rapid change in physical geography and in organisms, when new and higher forms appear which serve to draw the boundary line of new epochs. Nevertheless, geological history is a continuous progress; its periods and epochs shade into one another by imperceptible gradations, and all our subdivisions must needs be vague and more or less arbitrary.

HOW FOSSILS TELL OF THE GEOGRAPHY OF THE PAST. Fossils are used not only as a record of the development of life upon the earth, but also in testimony to the physical geography of past epochs. They indicate whether in any region the climate was tropical, temperate, or arctic. Since species spread slowly from some center of dispersion where they originate until some barrier limits their migration farther, the occurrence of the same species in rocks of the same system in different countries implies the absence of such barriers at the period. Thus in the collection of antarctic fossils referred to on page 294 there were shallow-water marine shells identical in species with Mesozoic shells found in India and in the southern extremity of South America. Since such organisms are not distributed by the currents of the deep sea and cannot migrate along its bottom, we infer a shallow-water connection in Mesozoic times between India, South America, and the antarctic region. Such a shallow-water connection would be offered along the marginal shelf of a continent uniting these now widely separated countries.

CHAPTER XV

THE PRE-CAMBRIAN SYSTEMS

THE EARTH'S BEGINNINGS. The geological record does not tell us of the beginnings of the earth. The history of the planet, as we have every reason to believe, stretches far back beyond the period of the oldest stratified rocks, and is involved in the history of the solar system and of the nebula,--the cloud of glowing gases or of cosmic dust,--from which the sun and planets are believed to have been derived.

THE NEBULAR HYPOTHESIS. It is possible that the earth began as a vaporous, shining sphere, formed by the gathering together of the material of a gaseous ring which had been detached from a cooling and shrinking nebula. Such a vaporous sphere would condense to a liquid, fiery globe, whose surface would become cold and solid, while the interior would long remain intensely hot because of the slow conductivity of the crust. Under these conditions the primeval atmosphere of the earth must have contained in vapor the water now belonging to the earth's crust and surface. It held also all the oxygen since locked up in rocks by their oxidation, and all the carbon dioxide which has since been laid away in limestones, besides that corresponding to the carbon of carbonaceous deposits, such as peat, coal, and petroleum. On this hypothesis the original atmosphere was dense, dark, and noxious, and enormously heavier than the atmosphere at present.

THE ACCRETION HYPOTHESIS. On the other hand, it has been recently suggested that the earth may have grown to its present size by the gradual accretion of meteoritic masses. Such cold, stony bodies might have come together at so slow a rate that the heat caused by their impact would not raise sensibly the temperature of the growing planet. Thus the surface of the earth may never have been hot and luminous; but as the loose aggregation of stony masses grew larger and was more and more compressed by its own gravitation, the heat thus generated raised the interior to high temperatures, while from time to time molten rock was intruded among the loose, cold meteoritic masses of the crust and outpoured upon the surface.

It is supposed that the meteorites of which the earth was built brought to it, as meteorites do now, various gases shut up within their pores. As the heat of the interior increased, these gases transpired to the surface and formed the primitive atmosphere and hydrosphere. The atmosphere has therefore grown slowly from the smallest beginnings. Gases emitted from the interior in volcanic eruptions and in other ways have ever added to it, and are adding to it now. On the other hand, the atmosphere has constantly suffered loss, as it has been robbed of oxygen by the oxidation of rocks in weathering, and of carbon dioxide in the making of limestones and carbonaceous deposits.

While all hypotheses of the earth's beginnings are as yet unproved speculations, they serve to bring to mind one of the chief lessons which geology has to teach,--that the duration of the earth in time, like the extension of the universe in space, is vastly beyond the power of the human mind to realize. Behind the history recorded in the rocks, which stretches back for many million years, lies the long unrecorded history of the beginnings of the planet; and still farther in the abysses of the past are dimly seen the cycles of the evolution of the solar system and of the nebula which gave it birth.

We pass now from the dim realm of speculation to the earliest era of the recorded history of the earth, where some certain facts may be observed and some sure inferences from them may be drawn.

THE ARCHEAN.

The oldest known sedimentary strata, wherever they are exposed by uplift and erosion, are found to be involved with a mass of crystalline rocks which possesses the same characteristics in all parts of the world. It consists of foliated rocks, gneisses, and schists of various kinds, which have been cut with dikes and other intrusions of molten rock, and have been broken, crumpled, and crushed, and left in interlocking masses so confused that their true arrangement can usually be made out only with the greatest difficulty if at all. The condition of this body of crystalline rocks is due to the fact that they have suffered not only from the faultings, foldings, and igneous intrusions of their time, but necessarily, also, from those of all later geological ages.

At present three leading theories are held as to the origin of these basal crystalline rocks.

1. They are considered by perhaps the majority of the geologists who have studied them most carefully to be igneous rocks intruded in a molten state among the sedimentary rocks involved with them. In many localities this relation is proved by the phenomena of contact; but for the most part the deformations which the rocks have since suffered again and again have been sufficient to destroy such evidence if it ever existed.

2. An older view regards them as profoundly altered sedimentary strata, the most ancient of the earth.

3. According to a third theory they represent portions of the earth's original crust; not, indeed, its original surface, but deeper portions uncovered by erosion and afterwards mantled with sedimentary deposits. All these theories agree that the present foliated condition of these rocks is due to the intense metamorphism which they have suffered.

It is to this body of crystalline rocks and the stratified rocks involved with it, which form a very small proportion of its mass, that the term ARCHEAN

(Greek, ARCHE, beginning) is applied by many geologists.

THE ALGONKIAN

In some regions there rests unconformably on the Archean an immense body of stratified rocks, thousands and in places even scores of thousands of feet thick, known as the ALGONKIAN. Great unconformities divide it into well-defined systems, but as only the scantiest traces of fossils appear here and there among its strata, it is as yet impossible to correlate the formations of different regions and to give them names of more than local application. We will describe the Algonkian rocks of two typical areas.

THE GRAND CANYON OF THE COLORADO. We have already studied a very ancient peneplain whose edge is exposed to view deep on the walls of the Colorado Canyon. The formation of flat-lying sandstone which covers this buried land surface is proved by its fossils to belong to the Cambrian,--the earliest period of the Paleozoic era. The tilted rocks on whose upturned edges the Cambrian sandstone rests are far older, for the physical break which separates them from it records a time interval during which they were upheaved to mountainous ridges and worn down to a low plain. They are therefore classified as Algonkian. They comprise two immense series. The upper is more than five thousand feet thick and consists of shales and sandstones with some limestones. Separated from it by an unconformity which does not appear in Figure 207, the lower division, seven thousand feet thick, consists chiefly of massive reddish sandstones with seven or more sheets of lava interbedded. The lowest member is a basal conglomerate composed of pebbles derived from the erosion of the dark crumpled schists beneath,--schists which are supposed to be Archean. As shown in Figure 207, a strong unconformity parts the schists and the Algonkian. The floor on which the Algonkian rests is remarkably even, and here again is proved an interval of incalculable length, during which an ancient land mass of Archean rocks was baseleveled before it received the cover of the sediments of the later age.

THE LAKE SUPERIOR REGION. In eastern Canada an area of pre-Cambrian rocks, Archean and Algonkian, estimated at two million square miles, stretches from the Great Lakes and the St. Lawrence River northward to the confines of the continent, inclosing Hudson Bay in the arms of a gigantic U. This immense area, which we have already studied as the Laurentian peneplain, extends southward across the Canadian border into northern Minnesota, Wisconsin, and Michigan. The rocks of this area are known to be pre-Cambrian; for the Cambrian strata, wherever found, lie unconformably upon them.

The general relations of the formations of that portion of the area which lies about Lake Superior are shown in Figure 262. Great unconformities, UU' separate the Algonkian both from the Archean and from the Cambrian, and divide it into three distinct systems, --the LOWER HURONIAN, the UPPER HURONIAN, and the KEWEENAWAN. The Lower and the Upper Huronian consist in the main of old sea muds and sands and limy oozes now changed to gneisses, schists, marbles, quartzites, slates, and other metamorphic rocks. The Keweenawan is composed of immense piles of lava, such as those of Iceland, overlain by bedded sandstones. What remains of these rock systems after the denudation of all later geologic ages is enormous. The Lower Huronian is more than a mile thick, the Upper Huronian more than two miles thick, while the Keweenawan exceeds nine miles in thickness. The vast length of Algonkian time is shown by the thickness of its marine deposits and by the cycles of erosion which it includes. In Figure 262 the student may read an outline of the history of the Lake Superior region, the deformations which it suffered, their relative severity, the times when they occurred, and the erosion cycles marked by the successive unconformities.

OTHER PRE-CAMBRIAN AREAS IN NORTH AMERICA. Pre-Cambrian rocks are exposed in various parts of the continent, usually by the erosion of mountain ranges in which their strata were infolded. Large areas occur in the maritime provinces of Canada. The core of the Green Mountains of Vermont is pre-Cambrian, and rocks of these systems occur in scattered patches in western Massachusetts. Here belong also the oldest rocks of the Highlands of the Hudson and of New Jersey. The Adirondack region, an outlier of the Laurentian region, exposes pre-Cambrian rocks, which have been metamorphosed and tilted by the intrusion of a great boss of igneous rock out of which the central peaks are carved. The core of the Blue Ridge and probably much of the Piedmont Belt are of this age. In the Black Hills the irruption of an immense mass of granite has caused or accompanied the upheaval of pre-Cambrian strata and metamorphosed them by heat and pressure into gneisses, schists, quartzites, and slates. In most of these mountainous regions the lowest strata are profoundly changed by metamorphism, and they can be assigned to the pre-Cambrian only where they are clearly overlain unconformably by formations proved to be Cambrian by their fossils. In the Belt Mountains of Montana, however, the Cambrian is underlain by Algonkian sediments twelve thousand feet thick, and but little altered.

MINERAL WEALTH OF THE PRE-CAMBRIAN ROCKS. The pre-Cambrian rocks are of very great economic importance, because of their extensive metamorphism and the enormous masses of igneous rock which they involve. In many parts of the country they are the source of supply of granite, gneiss, marble, slate, and other such building materials. Still more valuable are the stores of iron and copper and other metals which they contain.

At the present time the pre-Cambrian region about Lake Superior leads the world in the production of iron ore, its output for 1903 being more than five sevenths of the entire output of the whole United States, and exceeding that of any foreign country. The ore bodies consist chiefly of the red oxide of iron (hematite) and occur in troughs of the strata, underlain by some impervious rock. A theory held by many refers the ultimate source of the iron to the igneous rocks of the Archean. When these rocks were upheaved and subjected to weathering, their iron compounds were decomposed. Their iron was leached out and carried away to be laid in the Algonkian water bodies in beds of iron carbonate and other iron compounds. During the later ages, after the Algonkian strata had been uplifted to form part of the continent, a second concentration has taken place. Descending underground waters charged with oxygen have decomposed the iron carbonate and deposited the iron, in the form of iron oxide, in troughs of the strata where their downward progress was arrested by impervious floors.

The pre-Cambrian rocks of the eastern United States also are rich in iron. In certain districts, as in the Highlands of New Jersey, the black oxide of iron (magnetite) is so abundant in beds and disseminated grains that the ordinary surveyor's compass is useless.

The pre-Cambrian copper mines of the Lake Superior region are among the richest on the globe. In the igneous rocks copper, next to iron, is the most common of all the useful metals, and it was especially abundant in the Keweenawan lavas. After the Keweenawan was uplifted to form land, percolating waters leached out much of the copper diffused in the lava sheets and deposited it within steam blebs as amygdules of native copper, in cracks and fissures, and especially as a cement, or matrix, in the interbedded gravels which formed the chief aquifers of the region. The famous Calumet and Hecla mine follows down the dip of the strata to the depth of nearly a mile and works such an ancient conglomerate whose matrix is pure copper.

THE APPEARANCE OF LIFE. Sometime during the dim ages preceding the Cambrian, whether in the Archean or in the Algonkian we know not, occurred one of the most important events in the history of the earth. Life appeared for the first time upon the planet. Geology has no evidence whatever to offer as to whence or how life came. All analogies lead us to believe that its appearance must have been sudden. Its earliest forms are unknown, but analogy suggests that as every living creature has developed from a single cell, so the earliest organisms upon the globe--the germs from which all later life is supposed to have been evolved--were tiny, unicellular masses of protoplasm, resembling the amoeba of to-day in the simplicity of their structure.

Such lowly forms were destitute of any hard parts and could leave no evidence of their existence in the record of the rocks. And of their supposed descendants we find so few traces in the pre- Cambrian strata that the first steps in organic evolution must be supplied from such analogies in embryology as the following. The fertilized ovum, the cell with which each animal begins its life, grows and multiplies by cell division, and develops into a hollow globe of cells called the BLASTOSPHERE. This stage is succeeded by the stage of the GASTRULA,--an ovoid or cup-shaped body with a double wall of cells inclosing a body cavity, and with an opening, the primitive mouth. Each of these early embryological stages is represented by living animals,--the undivided cell by the PROTOZOA, the blastosphere by some rare forms, and the gastrula in the essential structure of the COELENTERATES,--the subkingdom to which the fresh-water hydra and the corals belong. All forms of animal life, from the coelenterates to the mammals, follow the same path in their embryological development as far as the gastrula stage, but here their paths widely diverge, those of each subkingdom going their own separate ways.

We may infer, therefore, that during the pre-Cambrian periods organic evolution followed the lines thus dimly traced. The earliest one-celled protozoa were probably succeeded by many-celled animals of the type of the blastosphere, and these by gastrula-like organisms. From the gastrula type the higher sub-divisions of animal life probably diverged, as separate branches from a common trunk. Much or all of this vast differentiation was accomplished before the opening of the next era; for all the subkingdoms are represented in the Cambrian except the vertebrates.

EVIDENCES OF PRE-CAMBRIAN LIFE. An indirect evidence of life during the pre-Cambrian periods is found in the abundant and varied fauna of the next period; for, if the theory of evolution is correct, the differentiation of the Cambrian fauna was a long process which might well have required for its accomplishment a large part of pre-Cambrian time.

Other indirect evidences are the pre-Cambrian limestones, iron ores, and graphite deposits, since such minerals and rocks have been formed in later times by the help of organisms. If the carbonate of lime of the Algonkian limestones and marbles was extracted from sea water by organisms, as is done at present by corals, mollusks, and other humble animals and plants, the life of those ancient seas must have been abundant. Graphite, a soft black mineral composed of carbon and used in the manufacture of lead pencils and as a lubricant, occurs widely in the metamorphic pre-Cambrian rocks. It is known to be produced in some cases by the metamorphism of coal, which itself is formed of decomposed vegetal tissues. Seams of graphite may therefore represent accumulations of vegetal matter such as seaweed. But limestone, iron ores, and graphite can be produced by chemical processes, and their presence in the pre-Cambrian makes it only probable, and not certain, that life existed at that time.

PRE-CAMBRIAN FOSSILS. Very rarely has any clear trace of an organism been found in the most ancient chapters of the geological record, so many of their leaves have been destroyed and so far have their pages been defaced. Omitting structures whose organic nature has been questioned, there are left to mention a tiny seashell of one of the most lowly types,--a DISCINA from the pre-Cambrian rocks of the Colorado Canyon,--and from the pre-Cambrian rocks of Montana trails of annelid worms and casts of their burrows in ancient beaches, and fragments of the tests of crustaceans. These diverse forms indicate that before the Algonkian had closed, life was abundant and had widely differentiated. We may expect that other forms will be discovered as the rocks are closely searched.

PRE-CAMBRIAN GEOGRAPHY. Our knowledge is far too meager to warrant an attempt to draw the varying outlines of sea and land during the Archean and Algonkian eras. Pre-Cambrian time probably was longer than all later geological time down to the present, as we may infer from the vast thicknesses of its rocks and the unconformities which part them. We know that during its long periods land masses again and again rose from the sea, were worn low, and were submerged and covered with the waste of other lands. But the formations of separated regions cannot be correlated because of the absence of fossils, and nothing more can be made out than the detached chapters of local histories, such as the outline given of the district about Lake Superior.

The pre-Cambrian rocks show no evidence of any forces then at work upon the earth except the forces which are at work upon it now. The most ancient sediments known are so like the sediments now being laid that we may infer that they were formed under conditions essentially similar to those of the present time. There is no proof that the sands of the pre-Cambrian sandstones were swept by any more powerful waves and currents than are offshore sands to-day, or that the muds of the pre-Cambrian shales settled to the sea floor in less quiet water than such muds settle in at present. The pre-Cambrian lands were, no doubt, worn by wind and weather, beaten by rain, and furrowed by streams as now, and, as now, they fronted the ocean with beaches on which waves dashed and along which tidal currents ran.

Perhaps the chief difference between the pre-Cambrian and the present was the absence of life upon the land. So far as we have any knowledge, no forests covered the mountain sides, no verdure carpeted the plains, and no animals lived on the ground or in the air. It is permitted to think of the most ancient lands as deserts of barren rock and rock waste swept by rains and trenched by powerful streams. We may therefore suppose that the processes of their destruction went on more rapidly than at present.

CHAPTER XVI

THE CAMBRIAN

THE PALEOZOIC ERA. The second volume of the geological record, called the Paleozoic (Greek, PALAIOS, ancient; ZOE, life), has come down to us far less mutilated and defaced than has the first volume, which contains the traces of the most ancient life of the globe. Fossils are far more abundant in the Paleozoic than in the earlier strata, while the sediments in which they were entombed have suffered far less from metamorphism and other causes, and have been less widely buried from view, than the strata of the pre-Cambrian groups. By means of their fossils we can correlate the formations of widely separated regions from the beginning of the Paleozoic on, and can therefore trace some outline of the history of the continents.

Paleozoic time, although shorter than the pre-Cambrian as measured by the thickness of the strata, must still be reckoned in millions of years. During this vast reach of time the changes in organisms were very great. It is according to the successive stages in the advance of life that the Paleozoic formations are arranged in five systems,--the CAMBRIAN, the ORDOVICIAN, the SILURIAN, the DEVONIAN, and the CARBONIFEROUS. On the same basis the first three systems are grouped together as the older Paleozoic, because they alike are characterized by the dominance of the invertebrates; while the last two systems are united in the later Paleozoic, and are characterized, the one by the dominance of fishes, and the other by the appearance of amphibians and reptiles.

Each of these systems is world-wide in its distribution, and may be recognized on any continent by its own peculiar fauna. The names first given them in Great Britain have therefore come into general use, while their subdivisions, which often cannot be correlated in different countries and different regions, are usually given local names.

The first three systems were named from the fact that their strata are well displayed in Wales. The Cambrian carries the Roman name of Wales, and the Ordovician and Silurian the names of tribes of ancient Britons which inhabited the same country. The Devonian is named from the English county Devon, where its rocks were early studied. The Carboniferous was so called from the large amount of coal which it was found to contain in Great Britain and continental Europe.

THE CAMBRIAN

DISTRIBUTION OF STRATA. The Cambrian rocks outcrop in narrow belts about the pre-Cambrian areas of eastern Canada and the Lake Superior region, the Adirondacks and the Green Mountains. Strips of Cambrian formations occupy troughs in the pre-Cambrian rocks of New England and the maritime provinces of Canada; a long belt borders on the west the crystalline rocks of the Blue Ridge; and on the opposite side of the continent the Cambrian reappears in the mountains of the Great Basin and the Canadian Rockies. In the Mississippi valley it is exposed in small districts where uplift has permitted the stripping off of younger rocks. Although the areas of outcrop are small, we may infer that Cambrian rocks were widely deposited over the continent of North America.

PHYSICAL GEOGRAPHY. The Cambrian system of North America comprises three distinct series, the LOWER CAMBRIAN, the MIDDLE CAMBRIAN, and the UPPER CAMBRIAN, each of which is characterized by its own peculiar fauna. In sketching the outlines of the continent as it was at the beginning of the Paleozoic, it must be remembered that wherever the Lower Cambrian formations now are found was certainly then sea bottom, and wherever the Lower Cambrian are wanting, and the next formations rest directly on pre-Cambrian rocks, was probably then land.

EARLY CAMBRIAN GEOGRAPHY. In this way we know that at the opening of the Cambrian two long, narrow mediterranean seas stretched from north to south across the continent. The eastern sea extended from the Gulf of St. Lawrence down the Champlain-Hudson valley and thence along the western base of the Blue Ridge south at least to Alabama. The western sea stretched from the Canadian Rockies over the Great Basin and at least as far south as the Grand Canyon of the Colorado in Arizona.

Between these mediterraneans lay a great central land which included the pre-Cambrian U-shaped area of the Laurentian peneplain, and probably extended southward to the latitude of New Orleans. To the east lay a land which we may designate as APPALACHIA, whose western shore line was drawn along the site of the present Blue Ridge, but whose other limits are quite unknown. The land of Appalachia must have been large, for it furnished a great amount of waste during the entire Paleozoic era, and its eastern coast may possibly have lain even beyond the edge of the present continental shelf.

On the western side of the continent a narrow land occupied the site of the Sierra Nevada Mountains.

Thus, even at the beginning of the Paleozoic, the continental plateau of North America had already been left by crustal movements in relief above the abysses of the great oceans on either side. The mediterraneans which lay upon it were shallow, as their sediments prove. They were EPICONTINENTAL SEAS; that is, they rested UPON (Greek, EPI) the submerged portion of the continental plateau. We have no proof that the deep ocean ever occupied any part of where North America now is.

The Middle and Upper Cambrian strata are found together with the Lower Cambrian over the area of both the eastern and the western mediterraneans, so that here the sea continued during the entire period. The sediments throughout are those of shoal water. Coarse cross-bedded sandstones record the action of strong shifting currents which spread coarse waste near shore and winnowed it of finer stuff. Frequent ripple marks on the bedding planes of the strata prove that the loose sands of the sea floor were near enough to the surface to be agitated by waves and tidal currents. Sun cracks show that often the outgoing tide exposed large muddy flats to the drying action of the sun. The fossils, also, of the strata are of kinds related to those which now live in shallow waters near the shore.

The sediments which gathered in the mediterranean seas were very thick, reaching in places the enormous depth of ten thousand feet. Hence the bottoms of these seas were sinking troughs, ever filling with waste from the adjacent land as fast as they subsided.

LATE CAMBRIAN GEOGRAPHY. The formations of the Middle and Upper Cambrian are found resting unconformably on the pre-Cambrian rocks from New York westward into Minnesota and at various points in the interior, as in Missouri and in Texas. Hence after earlier Cambrian time the central land subsided, with much the same effect as if the Mississippi valley were now to lower gradually, and the Gulf of Mexico to spread northward until it entered Lake Superior. The Cambrian seas transgressed the central land and strewed far and wide behind their advancing beaches the sediments of the later Cambrian upon an eroded surface of pre-Cambrian rocks.

The succession of the Cambrian formations in North America records many minor oscillations and varying conditions of physical geography; yet on the whole it tells of widening seas and lowering lands. Basal conglomerates and coarse sandstones which must have been laid near shore are succeeded by shaly sandstones, sandy shales, and shales. Toward the top of the series heavy beds of limestone, extending from the Blue Ridge to Missouri, speak of clear water, and either of more distant shores or of neighboring lands which were worn or sunk so low that for the most part their waste was carried to the sea in solution.

In brief, the Cambrian was a period of submergence. It began with the larger part of North America emerged as great land masses. It closed with most of the interior of the continental plateau covered with a shallow sea.

THE LIFE OF THE CAMBRIAN PERIOD

It is now for the first time that we find preserved in the offshore deposits of the Cambrian seas enough remains of animal life to be properly called a fauna. Doubtless these remains are only the most fragmentary representation of the life of the time, for the Cambrian rocks are very old and have been widely metamorphosed. Yet the five hundred and more species already discovered embrace all the leading types of invertebrate life, and are so varied that we must believe that their lines of descent stretch far back into the pre-Cambrian past.

PLANTS. No remains of plants have been found in Cambrian strata, except some doubtful markings, as of seaweed.

SPONGES. The sponges, the lowest of the multicellular animals, were represented by several orders. Their fossils are recognized by the siliceous spicules, which, as in modern sponges, either were scattered through a mass of horny fibers or were connected in a flinty framework.

COELENTERATES. This subkingdom includes two classes of interest to the geologist,--the HYDROZOA, such as the fresh-water hydra and the jellyfish, and the CORALS. Both classes existed in the Cambrian.

The Hydrozoa were represented not only by jellyfish but also by the

GRAPTOLITE, which takes its name from a fancied resemblance of some of its forms to a quill pen. It was a composite animal with a horny framework, the individuals of the colony living in cells strung on one or both sides along a hollow stem, and communicating by means of a common flesh in this central tube. Some graptolites were straight, and some curved or spiral; some were single stemmed, and others consisted of several radial stems united. Graptolites occur but rarely in the Upper Cambrian. In the Ordovician and Silurian they are very plentiful, and at the close of the Silurian they pass out of existence, never to return.

CORALS are very rarely found in the Cambrian, and the description of their primitive types is postponed to later chapters treating of periods when they became more numerous.

ECHINODERMS. This subkingdom comprises at present such familiar forms as the crinoid, the starfish, and the sea urchin. The structure of echinoderms is radiate. Their integument is hardened with plates or particles of carbonate of lime.

Of the free echinoderms, such as the starfish and the sea urchin, the former has been found in the Cambrian rocks of Europe, but neither have so far been discovered in the strata of this period in North America. The stemmed and lower division of the echinoderms was represented by a primitive type, the CYSTOID, so called from its saclike form, A small globular or ovate "calyx" of calcareous plates, with an aperture at the top for the mouth, inclosed the body of the animal, and was attached to the sea bottom by a short flexible stalk consisting of disks of carbonate of lime held together by a central ligament.

ARTHOPODS. These segmented animals with "jointed feet," as their name suggests, may be divided in a general way into water breathers and air breathers. The first-named and lower division comprises the class of the CRUSTACEA,--arthropods protected by a hard exterior skeleton, or "crust,"-- of which crabs, crayfish, and lobsters are familiar examples. The higher division, that of the air breathers, includes the following classes: spiders, scorpions, centipedes, and insects.

THE TRILOBITE. The aquatic arthropods, the Crustacea, culminated before the air breathers; and while none of the latter are found in the Cambrian, the former were the dominant life of the time in numbers, in size, and in the variety of their forms. The leading crustacean type is the TRILOBITE, which takes its name from the three lobes into which its shell is divided longitudinally. There are also three cross divisions,--the head shield, the tail shield, and between the two the thorax, consisting of a number of distinct and unconsolidated segments. The head shield carries a pair of large, crescentic, compound eyes, like those of the insect. The eye varies greatly in the number of its lenses, ranging from fourteen in some species to fifteen thousand in others. Figure 268, C, is a restoration of the trilobite, and shows the appendages, which are found preserved only in the rarest cases.

During the long ages of the Cambrian the trilobite varied greatly. Again and again new species and genera appeared, while the older types became extinct. For this reason and because of their abundance, trilobites are used in the classification of the Cambrian system. The Lower Cambrian is characterized by the presence of a trilobitic fauna in which the genus Olenellus is predominant. This, the OLENELLUS ZONE, is one of the most important platforms in the entire geological series; for, the world over, it marks the beginning of Paleozoic time, while all underlying strata are classified as pre-Cambrian. The Middle Cambrian is marked by the genus Paradoxides, and the Upper Cambrian by the genus Olenus. Some of the Cambrian trilobites were giants, measuring as much as two feet long, while others were the smallest of their kind, a fraction of an inch in length.

Another type of crustacean which lived in the Cambrian and whose order is still living is illustrated in Figure 269.

WORMS. Trails and burrows of worms have been left on the sea beaches and mud flats of all geological times from the Algonkian to the present.

BRACHIOPODS. These soft-bodied animals, with bivalve shells and two interior armlike processes which served for breathing, appeared in the Algonkian, and had now become very abundant. The two valves of the brachiopod shell are unequal in size, and in each valve a line drawn from the beak to the base divides the valve into two equal parts. It may thus be told from the pelecypod mollusk, such as the clam, whose two valves are not far from equal in size, each being divided into unequal parts by a line dropped from the beak.

Brachiopods include two orders. In the most primitive order--that of the INARTICULATE brachiopods--the two valves are held together only by muscles of the animal, and the shell is horny or is composed of phosphate of lime. The DISCINA, which began in the Algonkian, is of this type, as is also the LINGULELLA of the Cambrian. Both of these genera have lived on during the millions of years of geological time since their introduction, handing down from generation to generation with hardly any change to their descendants now living off our shores the characters impressed upon them at the beginning.

The more highly organized ARTICULATE brachiopods have valves of carbonate of lime more securely joined by a hinge with teeth and sockets (Fig. 270). In the Cambrian the inarticulates predominate, though the articulates grow common toward the end of the period.

MOLLUSKS. The three chief classes of mollusks--the PELECYPODS (represented by the oyster and clam of to-day), the GASTROPODS (represented now by snails, conches, and periwinkles), and the CEPHALOPODS (such as the nautilus, cuttlefish, and squids)--were all represented in the Cambrian, although very sparingly.

Pteropods, a suborder of the gastropods, appeared in this age. Their papery shells of carbonate of lime are found in great numbers from this time on.

Cephalopods, the most highly organized of the mollusks, started into existence, so far as the record shows, toward, the end of the Cambrian, with the long extinct ORTHOCERAS (STRAIGHTHORN) and the allied genera of its family. The Orthoceras had a long, straight, and tapering shell, divided by cross partitions into chambers. The animal lived in the "body chamber" at the larger end, and walled off the other chambers from it in succession during the growth of the shell. A central tube, the SIPHUNCLE, passed through from the body chamber to the closed tip of the cone.

The seashells, both brachiopods and mollusks, are in some respects the most important to the geologist of all fossils. They have been so numerous, so widely distributed, and so well preserved because of their durable shells and their station in growing sediments, that better than any other group of organisms they can be used to correlate the strata of different regions and to mark by their slow changes the advance of geological time.

CLIMATE. The life of Cambrian times in different countries contains no suggestion of any marked climatic zones, and as in later periods a warm climate probably reached to the polar regions.

CHAPTER XVII

THE ORDOVICIAN AND SILURIAN [Footnote: Often known as the Lower Silurian.]

THE ORDOVICIAN

In North America the Ordovician rocks lie conformably on the Cambrian. The two periods, therefore, were not parted by any deformation, either of mountain making or of continental uplift. The general submergence which marked the Cambrian continued into the succeeding period with little interruption.

SUBDIVISIONS AND DISTRIBUTION OF STRATA. The Ordovician series, as they have been made out in New York, are given for reference in the following table, with the rocks of which they are chiefly composed:

5 Hudson . . . . shales 4 Utica . . . . shales 3 Trenton . . . . limestones 2 Chazy . . . . limestones 1 Calciferous . . . . . sandy limestones

These marine formations of the Ordovician outcrop about the Cambrian and pre-Cambrian areas, and, as borings show, extend far and wide over the interior of the continent beneath more recent strata. The Ordovician sea stretched from Appalachia across the Mississippi valley. It seems to have extended to California, although broken probably by several mountainous islands in the west.

PHYSICAL GEOGRAPHY. The physical history of the period is recorded in the succession of its formations. The sandstones of the Upper Cambrian, as we have learned, tell of a transgressing sea which gradually came to occupy the Mississippi valley and the interior of North America. The limestones of the early and middle Ordovician show that now the shore had become remote and the lands had become more low. The waters now had cleared. Colonies of brachiopods and other lime-secreting animals occupied the sea bottom, and their debris mantled it with sheets of limy ooze. The sandy limestones of the Calciferous record the transition stage from the Cambrian when some sand was still brought in from shore. The highly fossiliferous limestones of the Trenton tell of clear water and abundant life. We need not regard this epicontinental sea as deep. No abysmal deposits have been found, and the limestones of the period are those which would be laid in clear, warm water of moderate depth like that of modern coral seas.

The shales of the Utica and Hudson show that the waters of the sea now became clouded with mud washed in from land. Either the land was gradually uplifted, or perhaps there had arrived one of those periodic crises which, as we may imagine, have taken place whenever the crust of the shrinking earth has slowly given way over its great depressions, and the ocean has withdrawn its waters into deepening abysses. The land was thus left relatively higher and bordered with new coastal plains. The epicontinental sea was shoaled and narrowed, and muds were washed in from the adjacent lands.

THE TACONIC DEFORMATION. The Ordovician was closed by a deformation whose extent and severity are not yet known. From the St. Lawrence River to New York Bay, along the northwestern and western border of New England, lies a belt of Cambrian-Ordovician rocks more than a mile in total thickness, which accumulated during the long ages of those periods in a gradually subsiding trough between the Adirondacks and a pre-Cambrian range lying west of the Connecticut River. But since their deposition these ancient sediments have been crumpled and crushed, broken with great faults, and extensively metamorphosed. The limestones have recrystallized into marbles, among them the famous marbles of Vermont; the Cambrian sandstones have become quartzites, and the Hudson shale has been changed to a schist exposed on Manhattan Island and northward.

In part these changes occurred at the close of the Ordovician, for in several places beds of Silurian age rest unconformably on the upturned Ordovician strata; but recent investigations have made it probable that the crustal movements recurred at later times, and it was perhaps in the Devonian and at the close of the Carboniferous that the greater part of the deformation and metamorphism was accomplished. As a result of these movements,-- perhaps several times repeated,--a great mountain range was upridged, which has been long since leveled by erosion, but whose roots are now visible in the Taconic Mountains of western New England.

THE CINCINNATI ANTICLINE. Over an oval area in Ohio, Indiana, and Kentucky, whose longer axis extends from north to south through Cincinnati, the Ordovician strata rise in a very low, broad swell, called the Cincinnati anticline. The Silurian and Devonian strata thin out as they approach this area and seem never to have deposited upon it. We may regard it, therefore, as an island upwarped from the sea at the close of the Ordovician or shortly after.

PETROLEUM AND NATURAL GAS. These valuable illuminants and fuels are considered here because, although they are found in traces in older strata, it is in the Ordovician that they occur for the first time in large quantities. They range throughout later formations down to the most recent.

The oil horizons of California and Texas are Tertiary; those of Colorado, Cretaceous; those of West Virginia, Carboniferous; those of Pennsylvania, Kentucky, and Canada, Devonian; and the large field of Ohio and Indiana belongs to the Ordovician and higher systems.

Petroleum and natural gas, wherever found, have probably originated from the decay of organic matter when buried in sedimentary deposits, just as at present in swampy places the hydrogen and carbon of decaying vegetation combine to form marsh gas. The light and heat of these hydrocarbons we may think of, therefore, as a gift to the civilized life of our race from the humble organisms, both animal and vegetable, of the remote past, whose remains were entombed in the sediments of the Ordovician and later geological ages.

Petroleum is very widely disseminated throughout the stratified rocks. Certain limestones are visibly greasy with it, and others give off its characteristic fetid odor when struck with a hammer. Many shales are bituminous, and some are so highly charged that small flakes may be lighted like tapers, and several gallons of oil to the ton may be obtained by distillation.

But oil and gas are found in paying quantities only when certain conditions meet:

1. A SOURCE below, usually a bituminous shale, from whose organic matter they have been derived by slow change.

2. A RESERVOIR above, in which they have gathered. This is either a porous sandstone or a porous or creviced limestone.

3. Oil and gas are lighter than water, and are usually under pressure owing to artesian water. Hence, in order to hold them from escaping to the surface, the reservoir must have the shape of an ANTICLINE, DOME, or LENS.

4. It must also have an IMPERVIOUS COVER, usually a shale. In these reservoirs gas is under a pressure which is often enormous, reaching in extreme cases as high as a thousand five hundred pounds to the square inch. When tapped it rushes out with a deafening roar, sometimes flinging the heavy drill high in air. In accounting for this pressure we must remember that the gas has been compressed within the pores of the reservoir rock by artesian water, and in some cases also by its own expansive force. It is not uncommon for artesian water to rise in wells after the exhaustion of gas and oil.

LIFE OF THE ORDOVICIAN

During the ages of the Ordovician, life made great advances. Types already present branched widely into new genera and species, and new and higher types appeared.

Sponges continued from the Cambrian. Graptolites now reached their climax.

STROMATOPORA--colonies of minute hydrozoans allied to corals--grew in places on the sea floor, secreting stony masses composed of thin, close, concentric layers, connected by vertical rods. The Stromatopora are among the chief limestone builders of the Silurian and Devonian periods.

CORALS developed along several distinct lines, like modern corals they secreted a calcareous framework, in whose outer portions the polyps lived. In the Ordovician, corals were represented chiefly by the family of the CHOETETES, all species of which are long since extinct. The description of other types of corals will be given under the Silurian, where they first became abundant.

ECHINODERMS. The cystoid reaches its climax, but there appear now two higher types of echinoderms,--the crinoid and the starfish. The CRINOID, named from its resemblance to the lily, is like the cystoid in many respects, but has a longer stem and supports a crown of plumose arms. Stirring the water with these arms, it creates currents by which particles of food are wafted to its mouth. Crinoids are rare at the present time, but they grew in the greatest profusion in the warm Ordovician seas and for long ages thereafter. In many places the sea floor was beautiful with these graceful, flowerlike forms, as with fields of long-stemmed lilies. Of the higher, free-moving classes of the echinoderms, starfish are more numerous than in the Cambrian, and sea urchins make their appearance in rare archaic forms.

CRUSTACEANS. Trilobites now reach their greatest development and more than eleven hundred species have been described from the rocks of this period. It is interesting to note that in many species the segments of the thorax have now come to be so shaped that they move freely on one another. Unlike their Cambrian ancestors, many of the Ordovician trilobites could roll themselves into balls at the approach of danger. It is in this attitude, taken at the approach of death, that trilobites are often found in the Ordovician and later rocks. The gigantic crustaceans called the EURYPTERIDS were also present in this period.

The arthropods had now seized upon the land. Centipedes and insects of a low type, the earliest known land animals, have been discovered in strata of this system.

BRYOZOANS. No fossils are more common in the limestones of the time than the small branching stems and lacelike mats of the bryozoans,--the skeletons of colonies of a minute animal allied in structure to the brachiopod.

BRACHIOPODS. These multiplied greatly, and in places their shells formed thick beds of coquina. They still greatly surpassed the mollusks in numbers.

CEPHALOPODS. Among the mollusks we must note the evolution of the cephalopods. The primitive straight Orthoceras has now become abundant. But in addition to this ancestral type there appears a succession of forms more and more curved and closely coiled, as illustrated in Figure 285. The nautilus, which began its course in this period, crawls on the bottom of our present seas.

VERTEBRATES. The most important record of the Ordovician is that of the appearance of a new and higher type, with possibilities of development lying hidden in its structure that the mollusk and the insect could never hope to reach. Scales and plates of minute fishes found in the Ordovician rocks near Canon City, Colorado, show that the humblest of the vertebrates had already made its appearance. But it is probable that vertebrates had been on the earth for ages before this in lowly types, which, being destitute of hard parts, would leave no record.

THE SILURIAN

The narrowing of the seas and the emergence of the lands which characterized the closing epoch of the Ordovician in eastern North America continue into the succeeding period of the Silurian. New species appear and many old species now become extinct.

THE APPALACHIAN REGION. Where the Silurian system is most fully developed, from New York southward along the Appalachian Mountains, it comprises four series:

4 Salina . . . shales, impure limestones, gypsum, salt 3 Niagara . . . chiefly limestones 2 Clinton . . . sandstones, shales, with some limestones 1 Medina . . . conglomerates, sandstones

The rocks of these series are shallow-water deposits and reach the total thickness of some five thousand feet. Evidently they were laid over an area which was on the whole gradually subsiding, although with various gentle oscillations which are recorded in the different formations. The coarse sands of the heavy Medina formations record a period of uplift of the oldland of Appalachia, when erosion went on rapidly and coarse waste in abundance was brought down from the hills by swift streams and spread by the waves in wide, sandy flats. As the lands were worn lower the waste became finer, and during an epoch of transition--the Clinton-- there were deposited various formations of sandstones, shales, and limestones. The Niagara limestones testify to a long epoch of repose, when low-lying lands sent little waste down to the sea.

The gypsum and salt deposits of the Salina show that toward the close of the Silurian period a slight oscillation brought the sea floor nearer to the surface, and at the north cut off extensive tracts from the interior sea. In these wide lagoons, which now and then regained access to the open sea and obtained new supplies of salt water, beds of salt and gypsum were deposited as the briny waters became concentrated by evaporation under a desert climate. Along with these beds there were also laid shales and impure limestones.

In New York the "salt pans" of the Salina extended over an area one hundred and fifty miles long from east to west and sixty miles wide, and similar salt marshes occurred as far west as Cleveland, Ohio, and Goderich on Lake Huron. At Ithaca, New York, the series is fifteen hundred feet thick, and is buried beneath an equal thickness of later strata. It includes two hundred and fifty feet of solid salt, in several distinct beds, each sealed within the shales of the series.

Would you expect to find ancient beds of rock salt inclosed in beds of pervious sandstone?

The salt beds of the Salina are of great value. They are reached by well borings, and their brines are evaporated by solar heat and by boiling. The rock salt is also mined from deep shafts.

Similar deposits of salt, formed under like conditions, occur in the rocks of later systems down to the present. The salt beds of Texas are Permian, those of Kansas are Permian, and those of Louisiana are Tertiary.

THE MISSISSIPPI VALLEY. The heavy near-shore formations of the Silurian in the Appalachian region thin out toward the west. The Medina and the Clinton sandstones are not found west of Ohio, where the first passes into a shale and the second into a limestone. The Niagara limestone, however, spreads from the Hudson River to beyond the Mississippi, a distance of more than a thousand miles. During the Silurian period the Mississippi valley region was covered with a quiet, shallow, limestone-making sea, which received little waste from the low lands which bordered it.

The probable distribution of land and sea in eastern North America and western Europe is shown in Figure 287. The fauna of the interior region and of eastern Canada are closely allied with that of western Europe, and several species are identical. We can hardly account for this except by a shallow-water connection between the two ancient epicontinental seas. It was perhaps along the coastal shelves of a northern land connecting America and Europe by way of Greenland and Iceland that the migration took place, so that the same species came to live in Iowa and in Sweden.

THE WESTERN UNITED STATES. So little is found of the rocks of the system west of the Missouri River that it is quite probable that the western part of the United States had for the most part emerged from the sea at the close of the Ordovician and remained land during the Silurian. At the same time the western land was perhaps connected with the eastern land of Appalachia across Arkansas and Mississippi; for toward the south the Silurian sediments indicate an approach to shore.

LIFE OF THE SILURIAN

In this brief sketch it is quite impossible to relate the many changes of species and genera during the Silurian.

CORALS. Some of the more common types are familiarly known as cup corals, honeycomb corals, and chain corals. In the CUP CORALS the most important feature is the development of radiating vertical partitions, or SEPTA, in the cell of the polyp. Some of the cup corals grew in hemispherical colonies (Fig. 288), while many were separate individuals (Fig. 289), building a single conical, or horn-shaped cell, which sometimes reached the extreme size of a foot in length and two or three inches in diameter.

HONEYCOMB CORALS consist of masses of small, close-set prismatic cells, each crossed by horizontal partitions, or TABULAE, while the septa are rudimentary, being represented by faintly projecting ridges or rows of spines.

CHAIN CORALS are also marked by tabulae. Their cells form elliptical tubes, touching each other at the edges, and appearing in cross section like the links of a chain. They became extinct at the end of the Silurian.

The corals of the SYRINGOPORA family are similar in structure to chain corals, but the tubular columns are connected only in places.

To the echinoderms there is now added the BLASTOID (bud-shaped). The blastoid is stemmed and armless, and its globular "head" or "calyx," with its five petal-like divisions, resembles a flower bud. The blastoids became more abundant in the Devonian, culminated in the Carboniferous, and disappeared at the end of the Paleozoic.

The great eurypterids--some of which were five or six feet in length--and the cephalopods were still masters of the seas. Fishes were as yet few and small; trilobites and graptolites had now passed their prime and had diminished greatly in numbers. Scorpions are found in this period both in Europe and in America. The limestone-making seas of the Silurian swarmed with corals, crinoids, and brachiopods.

With the end of the Silurian period the AGE OF INVERTEBRATES comes to a close, giving place to the Devonian, the AGE OF FISHES.

CHAPTER XVIII

THE DEVONIAN

In America the Silurian is not separated from the Devonian by any mountain-making deformation or continental uplift. The one period passed quietly into the other. Their conformable systems are so closely related, and the change in their faunas is so gradual, that geologists are not agreed as to the precise horizon which divides them.

SUBDIVISIONS AND PHYSICAL GEOGRAPHY. The Devonian is represented in New York and southward by the following five series. We add the rocks of which they are chiefly composed.

5 Chemung . . . . . . sandstones and sandy shales 4 Hamilton . . . . . . shales and sandstones 3 Corniferous . . . . . . limestones 2 Oriskany . . . . . . sandstones 1 Helderberg . . . . . . limestones

The Helderberg is a transition epoch referred by some geologists to the Silurian. The thin sandstones of the Oriskany mark an epoch when waves worked over the deposits of former coastal plains. The limestones of the

Corniferous testify to a warm and clear wide sea which extended from the Hudson to beyond the Mississippi. Corals throve luxuriantly, and their remains, with those of mollusks and other lime-secreting animals, built up great beds of limestone. The bordering continents, as during the later Silurian, must now have been monotonous lowlands which sent down little of even the finest waste to the sea.

In the Hamilton the clear seas of the previous epoch became clouded with mud. The immense deposits of coarse sandstones and sandy shales of the Chemung, which are found off what was at the time the west coast of Appalachia, prove an uplift of that ancient continent.

The Chemung series extends from the Catskill Mountains to northeastern Ohio and south to northeastern Tennessee, covering an area of not less than a hundred thousand square miles. In eastern New York it attains three thousand feet in thickness; in Pennsylvania it reaches the enormous thickness of two miles; but it rapidly thins to the west. Everywhere the Chemung is made of thin beds of rapidly alternating coarse and fine sands and clays, with an occasional pebble layer, and hence is a shallow-water deposit. The fine material has not been thoroughly winnowed from the coarse by the long action of strong waves and tides. The sands and clays have undergone little more sorting than is done by rivers. We must regard the Chemung sandstones as deposits made at the mouths of swift, turbid rivers in such great amount that they could be little sorted and distributed by waves.

Over considerable areas the Chemung sandstones bear little or no trace of the action of the sea. The Catskill Mountains, for example, have as their summit layers some three thousand feet of coarse red sandstones of this series, whose structure is that of river deposits, and whose few fossils are chiefly of freshwater types. The Chemung is therefore composed of delta deposits, more or less worked over by the sea. The bulk of the Chemung equals that of the Sierra Nevada Mountains. To furnish this immense volume of sediment a great mountain range, or highland, must have been upheaved where the Appalachian lowland long had been. To what height the Devonian mountains of Appalachia attained cannot be told from the volume of the sediments wasted from them, for they may have risen but little faster than they were worn down by denudation. We may infer from the character of the waste which they furnished to the Chemung shores that they did not reach an Alpine height. The grains of the Chemung sandstones are not those which would result from mechanical disintegration, as by frost on high mountain peaks, but are rather those which would be left from the long chemical decay of siliceous crystalline rocks; for the more soluble minerals are largely wanting. The red color of much of the deposits points to the same conclusion. Red residual clays accumulated on the mountain sides and upland summits, and were washed as ocherous silt to mingle with the delta sands. The iron- bearing igneous rocks of the oldland also contributed by their decay iron in solution to the rivers, to be deposited in films of iron oxide about the quartz grains of the Chemung sandstones, giving them their reddish tints.

LIFE OF THE DEVONIAN

PLANTS. The lands were probably clad with verdure during Silurian times, if not still earlier; for some rare remains of ferns and other lowly types of vegetation have been found in the strata of that system. But it is in the Devonian that we discover for the first time the remains of extensive and luxuriant forests. This rich flora reached its climax in the Carboniferous, and it will be more convenient to describe its varied types in the next chapter.

RHIZOCARPS. In the shales of the Devonian are found microscopic spores of rhizocarps in such countless numbers that their weight must be reckoned in hundreds of millions of tons. It would seem that these aquatic plants culminated in this period, and in widely distant portions of the earth swampy flats and shallow lagoons were filled with vegetation of this humble type, either growing from the bottom or floating free upon the surface. It is to the resinous spores of the rhizocarps that the petroleum and natural gas from Devonian rocks are largely due. The decomposition of the spores has made the shales highly bituminous, and the oil and gas have accumulated in the reservoirs of overlying porous sandstones.

INVERTEBRATES. We must pass over the ever-changing groups of the invertebrates with the briefest notice. Chain corals became extinct at the close of the Silurian, but other corals were extremely common in the Devonian seas. At many places corals formed thin reefs, as at Louisville, Kentucky, where the hardness of the reef rock is one of the causes of the Falls of the Ohio.

Sponges, echinoderms, brachiopods, and mollusks were abundant. The cephalopods take a new departure. So far in all their various forms, whether straight, as the Orthoceras, or curved, or close- coiled as in the nautilus, the septum, or partition dividing the chambers, met the inner shell along a simple line, like that of the rim of a saucer. There now begins a growth of the septum by which its edges become sharply corrugated, and the suture, or line of juncture of the septum and the shell, is thus angled. The group in which this growth of the septum takes place is called the GONIATITE (Greek GONIA, angle).

VERTEBRATES. It is with the greatest interest that we turn now to study the backboned animals of the Devonian; for they are believed to be the ancestors of the hosts of vertebrates which have since dominated the earth. Their rudimentary structures foreshadowed what their descendants were to be, and give some clue to the earliest vertebrates from which they sprang. Like those whose remains are found in the lower Paleozoic systems, all of these Devonian vertebrates were aquatic and go under the general designation of fishes.

The lowest in grade and nearest, perhaps, to the ancestral type of vertebrates, was the problematic creature, an inch or so long, of Figure 297. Note the circular mouth not supplied with jaws, the lack of paired fins, and the symmetric tail fin, with the column of cartilaginous, ringlike vertebrae running through it to the end. The animal is probably to be placed with the jawless lampreys and hags,--a group too low to be included among true fishes.

OSTRACODERMS. This archaic group, long since extinct, is also too lowly to rank among the true fishes, for its members have neither jaws nor paired fins. These small, fishlike forms were cased in front with bony plates developed in the skin and covered in the rear with scales. The vertebrae were not ossified, for no trace of them has been found.

DEVONIAN FISHES. The TRUE FISHES of the Devonian can best be understood by reference to their descendants now living. Modern fishes are divided into several groups: SHARKS and their allies; DIPNOANS; GANOIDS, such as the sturgeon and gar; and TELEOSTS,-- most common fishes, such as the perch and cod.

SHARKS. Of all groups of living fishes the sharks are the oldest and still retain most fully the embryonic characters of their Paleozoic ancestors. Such characters are the cartilaginous skeleton, and the separate gill slits with which the throat wall is pierced and which are arranged in line like the gill openings of the lamprey. The sharks of the Silurian and Devonian are known to us chiefly by their teeth and fin spines, for they were unprotected by scales or plates, and were devoid of a bony skeleton. Figure 299 is a restoration of an archaic shark from a somewhat higher horizon. Note the seven gill slits and the lappetlike paired fins. These fins seem to be remnants of the continuous fold of skin which, as embryology teaches, passed from fore to aft down each side of the primitive vertebrate.

Devonian sharks were comparatively small. They had not evolved into the ferocious monsters which were later to be masters of the seas.

DIPNOANS, OR LUNG FISHES. These are represented to-day by a few peculiar fishes and are distinguished by some high structures which ally them with amphibians. An air sac with cellular spaces is connected with the gullet and serves as a rudimentary lung. It corresponds with the swim bladder of most modern fishes, and appears to have had a common origin with it. We may conceive that the primordial fishes not only had gills used in breathing air dissolved in water, but also developed a saclike pouch off the gullet. This sac evolved along two distinct lines. On the line of the ancestry of most modern fishes its duct was closed and it became the swim bladder used in flotation and balancing. On another line of descent it was left open, air was swallowed into it, and it developed into the rudimentary lung of the dipnoans and into the more perfect lungs of the amphibians and other air- breathing vertebrates.

One of the ancient dipnoans is illustrated in Figure 300. Some of the members of this order were, like the ostracoderms, cased in armor, but their higher rank is shown by their powerful jaws and by other structures. Some of these armored fishes reached twenty- five feet in length and six feet across the head. They were the tyrants of the Devonian seas.

GANOIDS. These take their name from their enameled plates or scales of bone. The few genera now surviving are the descendants of the tribes which swarmed in the Devonian seas. A restoration of one of a leading order, the FRINGE-FINNED ganoids, is given in Figure 301. The side fins, which correspond to the limbs of the higher vertebrates, are quite unlike those of most modern fishes. Their rays, instead of radiating from a common base, fringe a central lobe which contains a cartilaginous axis. The teeth of the Devonian ganoids show a complicated folded structure.

GENERAL CHARACTERISTICS OF DEVONIAN FISHES. THE NOTOCHORD IS PERSISTENT. The notochord is a continuous rod of cartilage, or gristle, which in the embryological growth of vertebrate animals supports the spinal nerve cord before the formation of the vertebrae. In most modern fishes and in all higher vertebrates the notochord is gradually removed as the bodies of the vertebrae are formed about it; but in the Devonian fishes it persists through maturity and the vertebrae remain incomplete.

THE SKELETON IS CARTILAGINOUS. This also is an embryological characteristic. In the Devonian fishes the vertebrae, as well as the other parts of the skeleton, have not ossified, or changed to bone, but remain in their primitive cartilaginous condition.

THE TAIL FIN IS VERTEBRATED. The backbone runs through the fin and is fringed above and below with its vertical rays. In some fishes with vertebrated tail fins the fin is symmetric, and this seems to be the primitive type. In others the tail fin is unsymmetric: the backbone runs into the upper lobe, leaving the two lobes of unequal size. In most modern fishes (the teleosts) the tail fin is not vertebrated: the spinal column ends in a broad plate, to which the diverging fin rays are attached.

But along with these embryonic characters, which were common to all Devonian fishes, there were other structures in certain groups which foreshadowed the higher structures of the land vertebrates which were yet to come: air sacs which were to develop into lungs, and cartilaginous axes in the side fins which were a prophecy of limbs. The vertebrates had already advanced far enough to prove the superiority of their type of structure to all others. Their internal skeleton afforded the best attachment for muscles and enabled them to become the largest and most powerful creatures of the time. The central nervous system, with the predominance given to the ganglia at the fore end of the nerve cord,--the brain,-- already endowed them with greater energy than the invertebrates; and, still more important, these structures contained the possibility of development into the more highly organized land vertebrates which were to rule the earth.

TELEOSTS. The great group of fishes called the teleosts, or those with complete bony skeletons, to which most modern fishes belong, may be mentioned here, although in the Devonian they had not yet appeared. The teleosts are a highly specialized type, adapted most perfectly to their aquatic environment. Heavy armor has been discarded, and reliance is placed instead on swiftness. The skeleton is completely ossified and the notochord removed. The vertebrae have been economically withdrawn from the tail, and the cartilaginous axis of the side fins has been fotfoid unnecessary. The air sac has become a swim bladder. In this complete specialization they have long since lost the possibility of evolving into higher types.

It is interesting to note that the modern teleosts in their embryological growth pass through the stages which characterized the maturity of their Devonian ancestors; their skeleton is cartilaginous and their tail fin vertebrated.

CHAPTER XIX

THE CARBONIFEROUS

The Carboniferous system is so named from the large amount of coal which it contains. Other systems, from the Devonian on, are coal bearing also, but none so richly and to so wide an extent. Never before or since have the peculiar conditions been so favorable for the formation of extensive coal deposits.

With few exceptions the Carboniferous strata rest on those of the Devonian without any marked unconformity; the one period passed quietly into the other, with no great physical disturbances.

The Carboniferous includes three distinct series. The lower is called the MISSISSIPPIAN, from the outcrop of its formations along the Mississippi River in central and southern Illinois and the adjacent portions of Iowa and Missouri. The middle series is called the PENNSYLVANIAN (or Coal Measures), from its wide occurrence over Pennsylvania. The upper series is named the PERMIAN, from the province of Perm in Russia.

THE MISSISSIPPIAN SERIES. In the interior the Mississippian is composed chiefly of limestones, with some shales, which tell of a clear, warm, epicontinental sea swarming with crinoids, corals, and shells, and occasionally clouded with silt from the land.

In the eastern region, New York had been added by uplift to the Appalachian land which now was united to the northern area. From eastern Pennsylvania southward there were laid in a subsiding trough, first, thick sandstones (the Pocono sandstone), and later still heavier shales,--the two together reaching the thickness of four thousand feet and more. We infer a renewed uplift of Appalachia similar to that of the later epochs of the Devonian, but as much less in amount as the volume of sediments is smaller.

THE PENNSYLVANIAN SERIES

The Mississippian was brought to an end by a quiet oscillation which lifted large areas slightly above the sea, and the Pennsylvanian began with a movement in the opposite direction. The sea encroached on the new land, and spread far and wide a great basal conglomerate and coarse sandstones. On this ancient beach deposit a group of strata rests which we must now interpret. They consist of alternating shales and sandstones, with here and there a bed of limestone and an occasional seam of coal. A stratum of fire clay commonly underlies a coal seam, and there occur also beds of iron ore. We give a typical section of a very small portion of the series at a locality in Pennyslvania. Although some of the minor changes are omitted, the section shows the rapid alternation of the strata:

Feet 9 Sandstone and shale . . . . 25 8 Limestone . . . . 18 7 Sandstone . . . . 10 6 Coal . . . . 1-6 5 Shale . . . . 0-2 4 Sandstone . . . . 40 3 Limestone . . . . 10 2 Coal . . . . 5-12 1 Fire clay 3

This section shows more coal than is usual; on the whole, coal seams do not take up more than one foot in fifty of the Coal Measures. They vary also in thickness more than is seen in the section, some exceptional seams reaching the thickness of fifty feet.

HOW COAL WAS MADE.

1. Coal is of vegetable origin. Examined under the microscope even

anthracite, or hard coal, is seen to contain carbonized vegetal tissues. There are also all gradations connecting the hardest anthracite--through semibituminous coal, bituminous or soft coal, lignite (an imperfect coal in which sometimes woody fibers may be seen little changed)--with peat and decaying vegetable tissues. Coal is compressed and mineralized vegetal matter. Its varieties depend on the perfection to which the peculiar change called bituminization has been carried, and also, as shown in the table below, on the degree to which the volatile substances and water have escaped, and on the per cent of carbon remaining.

| | Peat | Lignite | Bituminous Coal | Anthracite |
	Dismal Swamp	Texas	Penn.	Penn.
Moisture	78.89	14.67	1.30	2.74
Volatile matter	13.84	37.32	20.87	4.25
Fixed carbon	6.49	41.07	67.20	81.51
Ash	0.78	6.69	8.80	10.87

2. The vegetable remains associated with coal are those of land plants.

3. Coal accumulated in the presence of water; for it is only when thus protected from the air that vegetal matter is preserved.

4. The vegetation of coal accumulated for the most part where it grew; it was not generally drifted and deposited by waves and currents. Commonly the fire clay beneath the seam is penetrated with roots, and the shale above is packed with leaves of ferns and other plants as beautifully pressed as in a herbarium. There often is associated with the seam a fossil forest, with the stumps, which are still standing where they grew, their spreading roots, and the soil beneath, all changed to stone. In the Nova Scotia field, out of seventy-six distinct coal seams, twenty are underlain by old forest grounds.

The presence of fire clay beneath a seam points in the same direction. Such underclays withstand intense heat and are used in making fire brick, because their alkalies have been removed by the long-continued growth of vegetation.

Fuel coal is also too pure to have been accumulated by driftage. In that case we should expect to find it mixed with mud, while in fact it often contains no more ash than the vegetal matter would furnish from which it has been compressed.

These conditions are fairly met in the great swamps of river plains and deltas and of coastal plains, such as the great Dismal Swamp, where thousands of generations of forests with their undergrowths contribute their stems and leaves to form thick beds of peat. A coal seam is a fossil peat bed.

GEOGRAPHICAL CONDITIONS DURING THE PENNSYLVANIAN. The Carboniferous peat swamps were of vast extent. A map of the Coal Measures (Fig. 260) shows that the coal marshes stretched, with various interruptions of higher ground and straits of open water, from eastern Pennsylvania into Alabama, Texas, and Kansas. Some individual coal beds may still be traced over a thousand square miles, despite the erosion which they have suffered. It taxes the imagination to conceive that the varied region included within these limits was for hundreds of thousands of years a marshy plain covered with tropical jungles such as that pictured in Figure 304.

On the basis that peat loses four fifths of its bulk in changing to coal, we may reckon the thickness of these ancient peat beds. Coal seams six and ten feet thick, which are not uncommon, represent peat beds thirty and fifty feet in thickness, while mammoth coal seams fifty feet thick have been compressed from peat beds two hundred and fifty feet deep.

At the same time, the thousands of feet of marine and freshwater sediments, with their repeated alternations of limestones, sandstones, and shales, in which the seams of coal occur, prove a slow subsidence, with many changes in its rate, with halts when the land was at a stillstand, and with occasional movements upward.

When subsidence was most rapid and long continued the sea encroached far and wide upon the lowlands and covered the coal swamps with sands and muds and limy oozes. When subsidence slackened or ceased the land gained on the sea. Bays were barred, and lagoons as they gradually filled with mud became marshes. River deltas pushed forward, burying with their silts the sunken peat beds of earlier centuries, and at the surface emerged in broad, swampy flats,--like those of the deltas of the Mississippi and the Ganges,-- which soon were covered with luxuriant forests. At times a gentle uplift brought to sea level great coastal plains, which for ages remained mantled with the jungle, their undeveloped drainage clogged with its debris, and were then again submerged.

PHYSICAL GEOGRAPHY OF THE SEVERAL REGIONS. THE ACADIAN REGION lay on the eastern side of the northern land, where now are New Brunswick and Nova Scotia, and was an immense river delta. Here river deposits rich in coal accumulated to a depth of sixteen thousand feet. The area of this coal field is estimated at about thirty-six thousand square miles.

THE APPALACHIAN REGION skirts the Appalachian oldland on the west from the southern boundary of New York to northern Alabama, extending west into eastern Ohio. The Cincinnati anticline was now a peninsula, and the broad gulf which had lain between it and Appalachia was transformed at the beginning of the Pennsylvanian into wide marshy plains, now sinking beneath the sea and now emerging from it. This area subsided during the Carboniferous period to a depth of nearly ten thousand feet.

THE CENTRAL REGION lay west of the peninsula of the Cincinnati anticline, and extended from Indiana west into eastern Nebraska, and from central Iowa and Illinois southward about the ancient island in Missouri and Arkansas into Oklahoma and Texas. On the north the subsidence in this area was comparatively slight, for the Carboniferous strata scarcely exceed two thousand feet in thickness. But in Arkansas and Indian Territory the downward movement amounted to four and five miles, as is proved by shoal water deposits of that immense thickness.

The coal fields of Indiana, and Illinois are now separated by erosion from those lying west of the Mississippi River. At the south the Appalachian land seems still to have stretched away to the west across Louisiana and Mississippi into Texas, and this westward extension formed the southern boundary of the coal marshes of the continent.

The three regions just mentioned include the chief Carboniferous coal fields of North America. Including a field in central Michigan evidently formed in an inclosed basin (Fig. 260), and one in Rhode Island, the total area of American coal fields has been reckoned at not less than two hundred thousand square miles. We can hardly estimate the value of these great stores of fossil fuel to an industrial civilization. The forests of the coal swamps accumulated in their woody tissues the energy which they received from the sun in light and heat, and it is this solar energy long stored in coal seams which now forms the world's chief source of power in manufacturing.

THE WESTERN AREA. On the Great Plains beyond the Missouri River the Carboniferous strata pass under those of more recent systems. Where they reappear, as about dissected mountain axes or on stripped plateaus, they consist wholly of marine deposits and are devoid of coal. The rich coal fields of the West are of later date.

On the whole the Carboniferous seems to have been a time of subsidence in the West. Throughout the period a sea covered the Great Basin and the plateaus of the Colorado River. At the time of the greatest depression the sites of the central chains of the Rockies were probably islands, but early in the period they may have been connected with the broad lands to the south and east. Thousands of feet of Carboniferous sediments were deposited where the Sierra Nevada Mountains now stand.

THE PERMIAN. As the Carboniferous period drew toward its close the sea gradually withdrew from the eastern part of the continent. Where the sea lingered in the deepest troughs, and where inclosed basins were cut off from it, the strata of the Permian were deposited. Such are found in New Brunswick, in Pennsylvania and West Virginia, in Texas, and in Kansas. In southwestern Kansas extensive Permian beds of rock salt and gypsum show that here lay great salt lakes in which these minerals were precipitated as their brines grew dense and dried away.

In the southern hemisphere the Permian deposits are so extraordinary that they deserve a brief notice, although we have so far omitted mention of the great events which characterized the evolution of other continents than our own. The Permian fauna- flora of Australia, India, South Africa, and the southern part of South America are so similar that the inference is a reasonable one that these widely separated regions were then connected together, probably as extensions of a great antarctic continent.

Interbedded with the Permian strata of the first three countries named are extensive and thick deposits of a peculiar nature which are clearly ancient ground moraines. Clays and sand, now hardened to firm rock, are inset with unsorted stones of all sizes, which often are faceted and scratched. Moreover, these bowlder clays rest on rock pavements which are polished and scored with glacial markings. Hence toward the close of the Paleozoic the southern lands of the eastern hemisphere were invaded by great glaciers or perhaps by ice sheets like that which now shrouds Greenland.

These Permian ground moraines are not the first traces of the work of glaciers met with in the geological record. Similar deposits prove glaciation in Norway succeeding the pre-Cambrian stage of elevation, and Cambrian glacial drift has recently been found in China.

THE APPALACHIAN DEFORMATION. We have seen that during Paleozoic times a long, narrow trough of the sea lay off the western coast of the ancient land of Appalachia, where now are the Appalachian Mountains. During the long ages of this era the trough gradually subsided, although with many stillstands and with occasional slight oscillations upward. Meanwhile the land lying to the east was gradually uplifted at varying rates and with long pauses. The waste of the rising land was constantly transferred to the sinking marginal sea bottom, and on the whole the trough was filled with sediments as rapidly as it subsided. The sea was thus kept shallow, and at times, especially toward the close of the era, much of the area was upbuilt or raised to low, marshy, coastal plains. When the Carboniferous was ended the waste which had been removed from the land and laid along its margin in the successive formations of the Paleozoic had reached a thickness of between thirty and forty thousand feet.

Both by sedimentation and by subsidence the trough had now become a belt of weakness in the crust of the earth. Here the crust was now made of layers to the depth of six or seven miles. In comparison with the massive crystalline rocks of Appalachia on the east, the layered rock of the trough was weak to resist lateral pressure, as a ream of sheets of paper is weak when compared with a solid board of the same thickness. It was weaker also than the region to the west, since there the sediments were much thinner. Besides, by the long-continued depression the strata of the trough had been bent from the flat-lying attitude in which they were laid to one in which they were less able to resist a horizontal thrust.

There now occurred one of the critical stages in the history of the planet, when the crust crumples under its own weight and shrinks down upon a nucleus which is diminishing in volume and no longer able to support it. Under slow but resistless pressure the strata of the Appalachian trough were thrust against the rigid land, and slowly, steadily bent into long folds whose axes ran northeast-southwest parallel to the ancient coast line. It was on the eastern side next the buttress of the land that the deformation was the greatest, and the folds most steep and close. In central Pennsylvania and West Virginia the folds were for the most part open. South of these states the folds were more closely appressed, the strata were much broken, and the great thrust faults were formed which have been described already. In eastern Pennsylvania seams of bituminous coal were altered to anthracite, while outside the region of strong deformation, as in western Pennyslvania, they remained unchanged. An important factor in the deformation was the massive limestones of the Cambrian-Ordovician. Because of these thick, resistant beds the rocks were bent into wide folds and sheared in places with great thrust faults. Had the strata been weak shales, an equal pressure would have crushed and mashed them.

Although the great earth folds were slowly raised, and no doubt eroded in their rising, they formed in all probability a range of lofty mountains, with a width of from fifty to a hundred and twenty-five miles, which stretched from New York to central Alabama.

From their bases lowlands extended westward to beyond the Missouri River. At the same time ranges were upridged out of thick Paleozoic sediments both in the Bay of Fundy region and in the Indian Territory. The eastern portion of the North American continent was now well-nigh complete.

The date of the Appalachian deformation is told in the usual way. The Carboniferous strata, nearly two miles thick, are all infolded in the Appalachian ridges, while the next deposits found in this region--those of the later portion of the first period (the Trias) of the succeeding era--rest unconformably on the worn edges of the Appalachian folded strata. The deformation therefore took place about the close of the Paleozoic. It seems to have begun in the Permian, in, eastern Pennsylvania,--for here the Permian strata are wanting,--and to have continued into the Trias, whose earlier formations are absent over all the area.

With this wide uplift the subsidence of the sea floor which had so long been general in eastern North America came to an end. Deposition now gave place to erosion. The sedimentary record of the Paleozoic was closed, and after an unknown lapse of time, here unrecorded, the annals of the succeeding era were written under changed conditions.

In western North America the closing stages of the Paleozoic were marked by important oscillations. The Great Basin, which had long been a mediterranean sea, was converted into land over western Utah and eastern Nevada, while the waves of the Pacific rolled across California and western Nevada.

The absence of tuffs and lavas among the Carboniferous strata of North America shows that here volcanic action was singularly wanting during the entire period. Even the Appalachian deformation was not accompanied by any volcanic outbursts.

LIFE OF THE CARBONIFEROUS

PLANTS. The gloomy forests and dense undergrowths of the Carboniferous jungles would appear unfamiliar to us could we see them as they grew, and even a botanist would find many of their forms perplexing and hard to classify. None of our modern trees would meet the eye. Plants with conspicuous flowers of fragrance and beauty were yet to come. Even mosses and grasses were still absent.

Tree ferns lifted their crowns of feathery fronds high in air on trunks of woody tissue; and lowly herbaceous ferns, some belonging to existing families, carpeted the ground. Many of the fernlike forms, however, have distinct affinities with the cycads, of which they may be the ancestors, and some bear seeds and must be classed as gymnosperms.

Dense thickets, like cane or bamboo brakes, were composed of thick clumps of CALAMITES, whose slender, jointed stems shot up to a height of forty feet, and at the joints bore slender branches set with whorls of leaves. These were close allies of the Equiseta or "horsetails," of the present; but they bore characteristics of higher classes in the woody structures of their stems.

There were also vast monotonous forests, composed chiefly of trees belonging to the lycopods, and whose nearest relatives to-day are the little club mosses of our eastern woods. Two families of lycopods deserve special mention,--the Lepidodendrons and the Sigillaria.

The LEPIDODENDRON, or "scale tree," was a gigantic club moss fifty and seventy-five feet high, spreading toward the top into stout branches, at whose ends were borne cone-shaped spore cases. The younger parts of the tree were clothed with stiff needle-shaped leaves, but elsewhere the trunk and branches were marked with scalelike scars, left by the fallen leaves, and arranged in spiral rows.

The SIGILLARIA, or "seal tree," was similar to the Lepidodendron, but its fluted trunk divided into even fewer branches, and was dotted with vertical rows of leaf scars, like the impressions of a seal.

Both Lepidodendron and Sigillaria were anchored by means of great cablelike underground stems, which ran to long distances through the marshy ground. The trunks of both trees had a thick woody rind, inclosing loose cellular tissue and a pith. Their hollow stumps, filled with sand and mud, are common in the Coal Measures, and in them one sometimes finds leaves and stems, land shells, and the bones of little reptiles of the time which made their home there.

It is important to note that some of these gigantic lycopods, which are classed with the CRYPTOGAMS, or flowerless plants, had pith and medullary rays dividing their cylinders into woody wedges. These characters connect them with the PHANEROGAMS, or flowering plants. Like so many of the organisms of the remote past, they were connecting types from which groups now widely separated have diverged.

Gymnosperms, akin to the cycads, were also present in the Carboniferous forests. Such were the different species of CORDAITES, trees pyramidal in shape, with strap-shaped leaves and nutlike fruit. Other gymnosperms were related to the yews, and it was by these that many of the fossil nuts found in the Coal Measures were borne. It is thought by some that the gymnosperms had their station on the drier plains and higher lands.

The Carboniferous jungles extended over parts of Europe and of Asia, as well as eastern North America, and reached from the equator to within nine degrees of the north pole. Even in these widely separated regions the genera and species of coal plants are close akin and often identical.

INVERTEBRATES. Among the echinoderms, crinoids are now exceedingly abundant, sea urchins are more plentiful, and sea cucumbers are found now for the first time. Trilobites are rapidly declining, and pass away forever with the close of the period. Eurypterids are common; stinging scorpions are abundant; and here occur the first-known spiders.

We have seen that the arthropods were the first of all animals to conquer the realm of the air, the earliest insects appearing in the Ordovician. Insects had now become exceedingly abundant, and the Carboniferous forests swarmed with the ancestral types of dragon flies,--some with a spread of wing of more than two feet,-- May flies, crickets, and locusts. Cockroaches infested the swamps, and one hundred and thirty-three species of this ancient order have been discovered in the Carboniferous of North America. The higher flower-loving insects are still absent; the reign of the flowering plants has not yet begun. The Paleozoic insects were generalized types connecting the present orders. Their fore wings were still membranous and delicately veined, and used in flying; they had not yet become thick, and useful only as wing covers, as in many of their descendants.

FISHES still held to the Devonian types, with the exception that the strange ostracoderms now had perished.

AMPHIBIANS. The vertebrates had now followed the arthropods and the mollusks upon the land, and had evolved a higher type adapted to the new environment. Amphibians--the class to which frogs and salamanders belong--now appear, with lungs for breathing air and with limbs for locomotion on the land. Most of the Carboniferous amphibians were shaped like the salamander, with weak limbs adapted more for crawling than for carrying the body well above the ground. Some legless, degenerate forms were snakelike in shape.

The earliest amphibians differ from those of to-day in a number of respects. They were connecting types linking together fishes, from which they were descended, with reptiles, of which they were the ancestors. They retained the evidence of their close relationship with the Devonian fishes in their cold blood, their gills and aquatic habit during their larval stage, their teeth with dentine infolded like those of the Devonian ganoids but still more intricately, and their biconcave vertebrae which never completely ossified. These, the highest vertebrates of the time, had not yet advanced beyond the embryonic stage of the more or less cartilaginous skeleton and the persistent notochord.

On the other hand, the skull of the Carboniferous amphibians was made of close-set bony plates, like the skull of the reptile, rather than like that of the frog, with its open spaces (Figs. 313 and 314). Unlike modern amphibians, with their slimy skin, the Carboniferous amphibians wore an armor of bony scales over the ventral surface and sometimes over the back as well.

It is interesting to notice from the footprints and skeletons of these earliest-known vertebrates of the land what was the primitive number of digits. The Carboniferous amphibians had five- toed feet, the primitive type of foot, from which their descendants of higher orders, with a smaller number of digits, have diverged.

The Carboniferous was the age of lycopods and amphibians, as the Devonian had been the age of rhizocarps and fishes.

LIFE OF THE PERMIAN. The close of the Paleozoic was, as we have seen, a time of marked physical changes. The upridging of the Appalachians had begun and a wide continental uplift--proved by the absence of Permian deposits over large areas where sedimentation had gone on before--opened new lands for settlement to hordes of air-breathing animals. Changes of climate compelled extensive migrations, and the fauna of different regions were thus brought into conflict. The Permian was a time of pronounced changes in plant and animal life, and a transitional period between two great eras. The somber forests of the earlier Carboniferous, with their gigantic club mosses, were now replaced by forests of cycads, tree ferns, and conifers. Even in the lower Permian the Lepidodendron and Sigillaria were very rare, and before the end of the epoch they and the Calamites also had become extinct. Gradually the antique types of the Paleozoic fauna died out, and in the Permian rocks are found the last survivors of the cystoid, the trilobite, and the eurypterid, and of many long-lived families of brachiopods, mollusks, and other invertebrates. The venerable Orthoceras and the Goniatite linger on through the epoch and into the first period of the succeeding era. Forerunners of the great ammonite family of cephalopod mollusks now appear. The antique forms of the earlier Carboniferous amphibians continue, but with many new genera and a marked increase in size.

A long forward step had now been taken in the evolution of the vertebrates. A new and higher type, the reptiles, had appeared, and in such numbers and variety are they found in the Permian strata that their advent may well have occurred in a still earlier epoch. It will be most convenient to describe the Permian reptiles along with their descendants of the Mesozoic.

CHAPTER XX

THE MESOZOIC

With the close of the Permian the world of animal and vegetable life had so changed that the line is drawn here which marks the end of the old order and the beginning of the new and separates the Paleozoic from the succeeding era,--the Mesozoic, the Middle Age of geological history. Although the Mesozoic era is shorter than the Paleozoic, as measured by the thickness of their strata, yet its duration must be reckoned in millions of years. Its predominant life features are the culmination and the beginning of the decline of reptiles, amphibians, cephalopod mollusks, and cycads, and the advent of marsupial mammals, birds, teleost fishes, and angiospermous plants. The leading events of the long ages of the era we can sketch only in the most summary way.

The Mesozoic comprises three systems,--the TRIASSIC, named from its threefold division in Germany; the JURASSIC, which is well displayed in the Jura Mountains; and the CRETACEOUS, which contains the extensive chalk (Latin, creta) deposits of Europe.

In eastern North America the Mesozoic rocks are much less important than the Paleozoic, for much of this portion of the continent was land during the Mesozoic era, and the area of the Mesozoic rocks is small. In western North America, on the other hand, the strata of the Mesozoic--and of the Cenozoic also--are widely spread. The Paleozoic rocks are buried quite generally from view except where the mountain makings and continental uplifts of the Mesozoic and Cenozoic have allowed profound erosion to bring them to light, as in deep canyons and about mountain axes. The record of many of the most important events in the development of the continent during the Mesozoic and Cenozoic eras is found in the rocks of our western states.

THE TRIASSIC AND JURASSIC

EASTERN NORTH AMERICA. The sedimentary record interrupted by the Appalachian deformation was not renewed in eastern North America until late in the Triassic. Hence during this long interval the land stood high, the coast was farther out than now, and over our Atlantic states geological time was recorded chiefly in erosion forms of hill and plain which have long since vanished. The area of the later Triassic rocks of this region, which take up again the geological record, is seen in the map of Figure 260. They lie on the upturned and eroded edges of the older rocks and occupy long troughs running for the most part parallel to the Atlantic coast. Evidently subsidence was in progress where these rocks were deposited. The eastern border of Appalachia was now depressed. The oldland was warping, and long belts of country lying parallel to the shore subsided, forming troughs in which thousands of feet of sediment now gathered.

These Triassic rocks, which are chiefly sandstones, hold no marine fossils, and hence were not laid in open arms of the sea. But their layers are often ripple-marked, and contain many tracks of reptiles, imprints of raindrops, and some fossil wood, while an occasional bed of shale is filled with the remains of fishes. We may conceive, then, of the Connecticut valley and the larger trough to the southwest as basins gradually sinking at a rate perhaps no faster than that of the New Jersey coast to-day, and as gradually aggraded by streams from the neighboring uplands. Their broad, sandy flats were overflowed by wandering streams, and when subsidence gained on deposition shallow lakes overspread the alluvial plains. Perhaps now and then the basins became long, brackish estuaries, whose low shores were swept by the incoming tide and were in turn left bare at its retreat to receive the rain prints of passing showers and the tracks of the troops of reptiles which inhabited these valleys.

The Triassic rocks are mainly red sandstones,--often feldspathic, or arkose, with some conglomerates and shales. Considering the large amount of feldspathic material in these rocks, do you infer that they were derived from the adjacent crystalline and metamorphic rocks of the oldland of Appalachia, or from the sedimentary Paleozoic rocks which had been folded into mountains during the Appalachian deformation? If from the former, was the drainage of the northern Appalachian mountain region then, as now, eastward and southeastward toward the Atlantic? The Triassic sandstones are voluminous, measuring at least a mile in thickness, and are largely of coarse waste. What do you infer as to the height of the lands from which the waste was shed, or the direction of the oscillation which they were then undergoing? In the southern basins, as about Richmond, Virginia, are valuable beds of coal; what was the physical geography of these areas when the coal was being formed?

Interbedded with the Triassic sandstones are contemporaneous lava beds which were fed from dikes. Volcanic action, which had been remarkably absent in eastern North America during Paleozoic times, was well-marked in connection with the warping now in progress. Thick intrusive sheets have also been driven in among the strata, as, for example, the sheet of the Palisades of the Hudson, described on page 269.

The present condition of the Triassic sandstones of the Connecticut valley is seen in Figure 315. Were the beds laid in their present attitude? What was the nature of the deformation which they have suffered? When did the intrusion of lava sheets take place relative to the deformation? What effect have these sheets on the present topography, and why? Assuming that the Triassic deformation went on more rapidly than denudation, what was its effect on the topography of the time? Are there any of its results remaining in the topography of to-day? Do the Triassic areas now stand higher or lower than the surrounding country, and why? How do the Triassic sandstones and shales compare in hardness with the igneous and metamorphic rocks about them? The Jurassic strata are wanting over the Triassic areas and over all of eastern North America. Was this region land or sea, an area of erosion or sedimentation, during the Jurassic period? In New Jersey, Pennsylvania, and farther southwest the lowest strata of the next period, the Cretaceous, rest on the eroded edges of the earlier rocks. The surface on which they lie is worn so even that we must believe that at the opening of the Cretaceous the oldland of Appalachia, including the Triassic areas, had been baseleveled at least near the coast. When, therefore, did the deformation of the Triassic rocks occur?

WESTERN NORTH AMERICA. Triassic strata infolded in the Sierra Nevada Mountains carry marine fossils and reach a thickness of nearly five thousand feet. California was then under water, and the site of the Sierra was a subsiding trough slowly filling with waste from the Great Basin land to the east.

Over a long belt which reaches from Wyoming across Colorado into New Mexico no Triassic sediments are found, nor is there any evidence that they were ever present; hence this area was high land suffering erosion during the Triassic. On each side of it, in eastern Colorado and about the Black Hills, in western Texas, in Utah, over the site of the Wasatch Mountains, and southward into Arizona over the plateaus trenched by the Colorado River, are large areas of Triassic rocks, sandstones chiefly, with some rock salt and gypsum. Fossils are very rare and none of them marine. Here, then, lay broad shallow lakes often salt, and warped basins, in which the waste of the adjacent uplands gathered. To this system belong the sandstones of the Garden of the Gods in Colorado, which later earth movements have upturned with the uplifted mountain flanks.

The Jurassic was marked with varied oscillations and wide changes in the outline of sea and land.

Jurassic shales of immense thickness--now metamorphosed into slates--are found infolded into the Sierra Nevada Mountains. Hence during Jurassic times the Sierra trough continued to subside, and enormous deposits of mud were washed into it from the land lying to the east. Contemporaneous lava flows interbedded with the strata show that volcanic action accompanied the downwarp, and that molten rock was driven upward through fissures in the crust and outspread over the sea floor in sheets of lava.

THE SIERRA DEFORMATION. Ever since the middle of the Silurian, the Sierra trough had been sinking, though no doubt with halts and interruptions, until it contained nearly twenty-five thousand feet of sediment. At the close of the Jurassic it yielded to lateral pressure and the vast pile of strata was crumpled and upheaved into towering mountains. The Mesozoic muds were hardened and squeezed into slates. The rocks were wrenched and broken, and underground waters began the work of filling their fissures with gold-bearing quartz, which was yet to wait millions of years before the arrival of man to mine it. Immense bodies of molten rock were intruded into the crust as it suffered deformation, and these appear in the large areas of granite which the later denudation of the range has brought to light.

The same movements probably uplifted the rocks of the Coast Range in a chain of islands. The whole western part of the continent was raised and its seas and lakes were for the most part drained away.

THE BRITISH ISLES. The Triassic strata of the British Isles are continental, and include breccia beds of cemented talus, deposits of salt and gypsum, and sandstones whose rounded and polished grains are those of the wind-blown sands of deserts. In Triassic times the British Isles were part of a desert extending over much of northwestern Europe.

THE CRETACEOUS

The third great system of the Mesozoic includes many formations, marine and continental, which record a long and complicated history marked by great oscillations of the crust and wide changes in the outlines of sea and land.

EARLY CRETACEOUS. In eastern North America the lowest Cretaceous series comprises fresh-water formations which are traced from Nantucket across Martha's Vineyard and Long Island, and through New Jersey southward into Georgia. They rest unconformably on the Triassic sandstones and the older rocks of the region. The Atlantic shore line was still farther out than now in the northern states. Again, as during the Triassic, a warping of the crust formed a long trough parallel to the coast and to the Appalachian ridges, but cut off from the sea; and here the continental deposits of the early Cretaceous were laid.

Along the Gulf of Mexico the same series was deposited under like conditions over the area known as the Mississippi embayment, reaching from Georgia northwestward into Tennessee and thence across into Arkansas and southward into Texas.

In the Southwest the subsidence continued until the transgressing sea covered most of Mexico and Texas and extended a gulf northward into Kansas. In its warm and quiet waters limestones accumulated to a depth of from one thousand to five thousand feet in Texas, and of more than ten thousand feet in Mexico. Meanwhile the lowlands, where the Great Plains are now, received continental deposits; coal swamps stretched from western Montana into British Columbia.

THE MIDDLE CRETACEOUS. This was a land epoch. The early Cretaceous sea retired from Texas and Mexico, for its sediments are overlain unconformably by formations of the Upper Cretaceous. So long was the time gap between the two series that no species found in the one occurs in the other.

THE UPPER CRETACEOUS. There now began one of the most remarkable events in all geological history,--the great Cretaceous subsidence. Its earlier warpings were recorded in continental deposits,--wide sheets of sandstone, shale, and some coal,--which were spread from Texas to British Columbia. These continental deposits are overlain by a succession of marine formations whose vast area is shown on the map, Figure 260. We may infer that as the depression of the continent continued the sea came in far and wide over the coast lands and the plains worn low during the previous epochs. Upper Cretaceous formations show that south of New England the waters of the Atlantic somewhat overlapped the crystalline rocks of the Piedmont Belt and spread their waste over the submerged coastal plain. The Gulf of Mexico again covered the Mississippi embayment, reaching as far north as southern Illinois, and extended over Texas.

A mediterranean sea now stretched from the Gulf to the arctic regions and from central Iowa to the eastern shore of the Great Basin land at about the longitude of Salt Lake City, the Colorado Mountains rising from it in a chain of islands. Along with minor oscillations there were laid in the interior sea various formations of sandstones, shales, and limestones, and from Kansas to South Dakota beds of white chalk show that the clear, warm waters swarmed at times with foraminiferal life whose disintegrating microscopic shells accumulated in this rare deposit.

At this epoch a wide sea, interrupted by various islands, stretched across Eurasia from Wales and western Spain to China, and spread southward over much of the Sahara. To the west its waters were clear and on its floor the crumbled remains of foraminifers gathered in heavy accumulations of calcareous ooze,-- the white chalk of France and England. Sea urchins were also abundant, and sponges contributed their spicules to form nodules of flint.

THE LARAMIE. The closing stage of the Cretaceous was marked in North America by a slow uplift of the land. As the interior sea gradually withdrew, the warping basins of its floor were filled with waste from the rising lands about them, and over this wide area there were spread continental deposits in fresh-water lakes like the Great Lakes of the present, in brackish estuaries, and in river plains, while occasional oscillations now and again let in the sea. There were vast marshes in which there accumulated the larger part of the valuable coal seams of the West. The Laramie is the coal-bearing series of the West, as the Pennsylvanian is of the eastern part of our country.

THE ROCKY MOUNTAIN DEFORMATION. At the close of the Cretaceous we enter upon an epoch of mountain-making far more extensive than any which the continent had witnessed. The long belt lying west of the ancient axes of the Colorado Islands and east of the Great Basin land had been an area of deposition for many ages, and in its subsiding troughs Paleozoic and Mesozoic sediments had gathered to the depth of many thousand feet. And now from Mexico well-nigh to the Arctic Ocean this belt yielded to lateral pressure. The Cretaceous limestones of Mexico were folded into lofty mountains. A massive range was upfolded where the Wasatch Mountains now are, and various ranges of the Rockies in Colorado and other states were upridged. However slowly these deformations were effected they were no doubt accompanied by world-shaking earthquakes, and it is known that volcanic eruptions took place on a magnificent scale. Outflows of lava occurred along the Wasatch, the laccoliths of the Henry Mountains were formed, while the great masses of igneous rock which constitute the cores of the Spanish Peaks and other western mountains were thrust up amid the strata. The high plateaus from which many of these ranges rise had not yet been uplifted, and the bases of the mountains probably stood near the level of the sea.

North America was now well-nigh completed. The mediterranean seas which so often had occupied the heart of the land were done away with, and the continent stretched unbroken from the foot of the Sierras on the west to the Fall Line of the Atlantic coastal plain on the east.

THE MESOZOIC PENEPLAIN. The immense thickness of the Mesozoic formations conveys to our minds some idea of the vast length of time involved in the slow progress of its successive ages. The same lesson is taught as plainly by the amount of denudation which the lands suffered during the era.

The beginning of the Mesozoic saw a system of lofty mountain ranges stretching from New York into central Alabama. The end of this long era found here a wide peneplain crossed by sluggish wandering rivers and overlooked by detached hills as yet unreduced to the general level. The Mesozoic era was long enough for the Appalachian Mountains, upridged at its beginning, to have been weathered and worn away and carried grain by grain to the sea. The same plain extended over southern New England. The Taconic range, uplifted partially at least at the close of the Ordovician, and the block mountains of the Triassic, together with the pre- Cambrian mountains of ancient Appalachia, had now all been worn to a common level with the Allegheny ranges. The Mesozoic peneplain has been upwarped by later crustal movements and has suffered profound erosion, but the remnants of it which remain on the upland of southern New England and the even summits of the Allegheny ridges suffice to prove that it once existed. The age of the Mesozoic peneplain is determined from the fact that the lower Tertiary sediments were deposited on its even surface when at the close of the era the peneplain was depressed along its edges beneath the sea.

LIFE OF THE MESOZOIC

PLANT LIFE OF THE TRIASSIC AND JURASSIC. The Carboniferous forests of lepidodendrons and sigillafids had now vanished from the earth. The uplands were clothed with conifers, like the Araucarian pines of South America and Australia. Dense forests of tree ferns throve in moist regions, and canebrakes of horsetails of modern type, but with stems reaching four inches in thickness, bordered the lagoons and marshes. Cycads were exceedingly abundant. These gymnosperms, related to the pines and spruces in structure and fruiting, but palmlike in their foliage, and uncoiling their long leaves after the manner of ferns, culminated in the Jurassic. From the view point of the botanist the Mesozoic is the Age of Cycads, and after this era they gradually decline to the small number of species now existing in tropical latitudes.

PLANT LIFE OF THE CRETACEOUS. In the Lower Cretaceous the woodlands continued of much the same type as during the Jurassic. The forerunners now appeared of the modern dicotyls (plants with two seed leaves), and in the Middle Cretaceous the monocotyledonous group of palms came in. Palms are so like cycads that we may regard them as the descendants of some cycad type.

In the UPPER CRETACEOUS, cycads become rare. The highest types of flowering plants gain a complete ascendency, and forests of modern aspect cover the continent from the Gulf of Mexico to the Arctic Ocean. Among the kinds of forest trees whose remains are found in the continental deposits of the Cretaceous are the magnolia, the myrtle, the laurel, the fig, the tulip tree, the chestnut, the oak, beech, elm, poplar, willow, birch, and maple. Forests of Eucalyptus grew along the coast of New England, and palms on the Pacific shores of British Columbia. Sequoias of many varieties ranged far into northern Canada. In northern Greenland there were luxuriant forests of magnolias, figs, and cycads; and a similar flora has been disinterred from the Cretaceous rocks of Alaska and Spitzbergen. Evidently the lands within the Arctic Circle enjoyed a warm and genial climate, as they had done during the Paleozoic. Greenland had the temperature of Cuba and southern Florida, and the time was yet far distant when it was to be wrapped in glacier ice.

INVERTEBRATES. During the long succession of the ages of the Mesozoic, with their vast geographical changes, there were many and great changes in organisms. Species were replaced again and again by others better fitted to the changing environment. During the Lower Cretaceous alone there were no less than six successive changes in the faunas which inhabited the limestone-making sea which then covered Texas. We shall disregard these changes for the most part in describing the life of the era, and shall confine our view to some of the most important advances made in the leading types.

Stromatopora have disappeared. Protozoans and sponges are exceedingly abundant, and all contribute to the making of Mesozoic strata. Corals have assumed a more modern type. Sea urchins have become plentiful; crinoids abound until the Cretaceous, where they begin their decline to their present humble station.

Trilobites and eurypterids are gone. Ten-footed crustaceans abound of the primitive long-tailed type (represented by the lobster and the crayfish), and in the Jurassic there appears the modern short- tailed type represented by the crabs. The latter type is higher in organization and now far more common. In its embryological development it passes through the long-tailed stage; connecting links in the Mesozoic also indicate that the younger type is the offshoot of the older.

Insects evolve along diverse lines, giving rise to beetles, ants, bees, and flies.

Brachiopods have dwindled greatly in the number of their species, while mollusks have correspondingly increased. The great oyster family dates from here.

Cephalopods are now to have their day. The archaic Orthoceras lingers on into the Triassic and becomes extinct, but a remarkable development is now at hand for the more highly organized descendants of this ancient line. We have noticed that in the Devonian the sutures of some of the chambered shells become angled, evolving the Goniatite type. The sutures now become lobed and corrugated in Ceratites. The process was carried still farther, and the sutures were elaborately frilled in the great order of the Ammonites. It was in the Jurassic that the Ammonites reached their height. No fossils are more abundant or characteristic of their age. Great banks of their shells formed beds of limestone in warm seas the world over.

The ammonite stem branched into a most luxuriant variety of forms. The typical form was closely coiled like a nautilus. In others the coil was more or less open, or even erected into a spiral. Some were hook-shaped, and there were members of the order in which the shell was straight, and yet retained all the internal structures of its kind. At the end of the Mesozoic the entire tribe of ammonites became extinct.

The Belemnite (Greek, belemnon, a dart) is a distinctly higher type of cephalopod which appeared in the Triassic, became numerous and varied in the Jurassic and Cretaceous, and died out early in the Tertiary. Like the squids and cuttlefish, of which it was the prototype, it had an internal calcareous shell. This consisted of a chambered and siphuncled cone, whose point was sheathed in a long solid guard somewhat like a dart. The animal carried an ink sac, and no doubt used it as that of the modern cuttlefish is used,--to darken the water and make easy an escape from foes. Belemnites have sometimes been sketched with fossil sepia, or india ink, from their own ink sacs. In the belemnites and their descendants, the squids and cuttlefish, the cephalopods made the radical change from external to the internal shell. They abandoned the defensive system of warfare and boldly took up the offensive. No doubt, like their descendants, the belemnites were exceedingly active and voracious creatures.

FISHES AND AMPHIBIANS. In the Triassic and Jurassic, little progress was made among the fishes, and the ganoid was still the leading type. In the Cretaceous the teleosts, or bony fishes, made their appearance, while ganoids declined toward their present subordinate place.

The amphibians culminated in the Triassic, some being formidable creatures as large as alligators. They were still of the primitive Paleozoic types. Their pygmy descendants of more modern types are not found until later, salamanders appearing first in the Cretaceous, and frogs at the beginning of the Cenozoic.

No remains of amphibians have been discovered in the Jurassic. Do you infer from this that there were none in existence at that time?

REPTILES OF THE MESOZOIC

The great order of Reptiles made its advent in the Permian, culminated in the Triassic and Jurassic, and began to decline in the Cretaceous. The advance from the amphibian to the reptile was a long forward step in the evolution of the vertebrates. In the reptile the vertebrate skeleton now became completely ossified. Gills were abandoned and breathing was by lungs alone. The development of the individual from the egg to maturity was uninterrupted by any metamorphosis, such as that of the frog when it passes from the tadpole stage. Yet in advancing from the amphibian to the reptile the evolution of the vertebrate was far from finished. The cold-blooded, clumsy and sluggish, small- brained and unintelligent reptile is as far inferior to the higher mammals, whose day was still to come, as it is superior to the amphibian and the fish.

The reptiles of the Permian, the earliest known, were much like lizards in form of body. Constituting a transition type between the amphibians on the one hand, and both the higher reptiles and the mammals on the other, they retained the archaic biconcave vertebra of the fish and in some cases the persistent notochord, while some of them, the theromorphs, possessed characters allying them with mammals. In these the skull was remarkably similar to that of the carnivores, or flesh-eating mammals, and the teeth, unlike the teeth of any later reptiles, were divisible into incisors, canines, and molars, as are the teeth of mammals.

At the opening of the Mesozoic era reptiles were the most highly organized and powerful of any animals on the earth. New ranges of continental extent were opened to them, food was abundant, the climate was congenial, and they now branched into very many diverse types which occupied and ruled all fields,--the land, the air, and the sea. The Mesozoic was the Age of Reptiles.

THE ANCESTRY OF SURVIVING REPTILIAN TYPES. We will consider first the evolution of the few reptilian types which have survived to the present.

Crocodiles, the highest of existing reptiles, are a very ancient order, dating back to the lower Jurassic, and traceable to earlier ancestral, generalized forms, from which sprang several other orders also.

Turtles and tortoises are not found until the early Jurassic, when they already possessed the peculiar characteristics which set them off so sharply from other reptiles. They seem to have lived at first in shallow water and in swamps, and it is not until after the end of the Mesozoic that some of the order became adapted to life on the land.

The largest of all known turtles, Archelon, whose home was the great interior Cretaceous sea, was fully a dozen feet in length and must have weighed at least two tons. The skull alone is a yard long.

Lizards and snakes do not appear until after the close of the Mesozoic, although their ancestral lines may be followed back into the Cretaceous.

We will now describe some of the highly specialized orders peculiar to the Mesozoic.

LAND REPTILES. The DINOSAURS (terrible reptiles) are an extremely varied order which were masters of the land from the late Trias until the close of the Mesozoic era. Some were far larger than elephants, some were as small as cats; some walked on all fours, some were bipedal; some fed on the luxuriant tropical foliage, and others on the flesh of weaker reptiles. They may be classed in three divisions,--the FLESH-EATING DINOSAURS, the REPTILE-FOOTED DINOSAURS, and the BEAKED DINOSAURS,--the latter two divisions being herbivorous.

The FLESH-EATING DINOSAURS are the oldest known division of the order, and their characteristics are shown in Figure 329. As a class, reptiles are egg layers (oviparous); but some of the flesh- eating dinosaurs are known to have been VIVIPAROUS, i.e. to have brought forth their young alive. This group was the longest-lived of any of the three, beginning in the Trias and continuing to the close of the Mesozoic era.

Contrast the small fore limbs, used only for grasping, with the powerful hind limbs on which the animal stalked about. Some of the species of this group seem to have been able to progress by leaping in kangaroo fashion. Notice the sharp claws, the ponderous tail, and the skull set at right angles with the spinal column. The limb bones are hollow. The ceratosaurs reached a length of some fifteen feet, and were not uncommon in Colorado and the western lands in Jurassic times.

The REPTILE-FOOTED DINOSAURS (Sauropoda) include some of the biggest brutes which ever trod the ground. One of the largest, whose remains are found entombed in the Jurassic rocks of Wyoming and Colorado, is shown in Figure 330.

Note the five digits on the hind feet, the quadrupedal gait, the enormous stretch of neck and tail, the small head aligned with the vertebral column. Diplodocus was fully sixty-five feet long and must have weighed about twenty tons. The thigh bones of the Sauropoda are the largest bones which ever grew. That of a genus allied to the Diplodocus measures six feet and eight inches, and the total length of the animal must have been not far from eighty feet, the largest land animal known.

The Sauropoda became extinct when their haunts along the rivers and lakes of the western plains of Jurassic times were invaded by the Cretaceous interior sea.

The BEAKED DINOSAURS(Predentata) were distinguished by a beak sheathed with horn carried in front of the tooth-set jaw, and used, we may imagine, in stripping the leaves and twigs of trees and shrubs. We may notice only two of the most interesting types.

STEGOSAURUS (plated reptile) takes its name from the double row of bony plates arranged along its back. The powerful tail was armed with long spines, and the thick skin was defended with irregular bits of bone implanted in it. The brain of the stegosaur was smaller than that of any land vertebrate, while in the sacrum the nerve canal was enlarged to ten times the capacity of the brain cavity of the skull. Despite their feeble wits, this well-armored family lived on through millions of years which intervened between their appearance, at the opening of the Jurassic, and the close of the Cretaceous, when they became extinct.

A less stupid brute than the stegosaur was TRICERATOPS, the dinosaur of the three horns,--one horn carried on the nose, and a massive pair set over the eyes. Note the enormous wedge-shaped skull, with its sharp beak, and the hood behind resembling a fireman's helmet. Triceratops was fully twenty-five feet long, and of twice the bulk of an elephant. The family appeared in the Upper Cretaceous and became extinct at its close. Their bones are found buried in the fresh-water deposits of the time from Colorado to Montana and eastward to the Dakotas.

MARINE REPTILES. In the ocean, reptiles occupied the place now held by the aquatic mammals, such as whales and dolphins, and their form and structure were similarly modified to suit their environment. In the Ichthyosaurus (fish reptile), for example, the body was fishlike in form, with short neck and large, pointed head (Fig. 333).

A powerful tail, whose flukes were set vertical, and the lower one of which was vertebrated, served as propeller, while a large dorsal fin was developed as a cutwater. The primitive biconcave vertebrae of the fish and of the early land vertebrates were retained, and the limbs degenerated into short paddles. The skin of the ichthyosaur was smooth like that of a whale, and its food was largely fish and cephalopods, as the fossil contents of its stomach prove.

These sea monsters disported along the Pacific shore over northern California in Triassic times, and the bones of immense members of the family occur in the Jurassic strata of Wyoming. Like whales and seals, the ichthyosaurs were descended from land vertebrates which had become adapted to a marine habitat.

PLESIOSAURS were another order which ranged throughout the Mesozoic.

Descended from small amphibious animals, they later included great marine reptiles, characterized in the typical genus by long neck, snakelike head, and immense paddles. They swam in the Cretaceous interior sea of western North America.

MOSASAURS belong to the same order as do snakes and lizards, and are an offshoot of the same ancestral line of land reptiles. These snakelike creatures--which measured as much as forty-five feet in length--abounded in the Cretaceous seas. They had large conical teeth, and their limbs had become stout paddles.

The lower jaw of the mosasaur was jointed; the quadrate bone, which in all reptiles connects the bone of the lower jaw with the skull, was movable, and as in snakes the lower jaw could be used in thrusting prey down the throat. The family became extinct at the end of the Mesozoic, and left no descendants. One may imitate the movement of the lower jaw of the mosasaur by extending the arms, clasping the hands, and bending the elbows.

FLYING REPTILES. The atmosphere, which had hitherto been tenanted only by insects, was first conquered by the vertebrates in the Mesozoic. Pterosaurs, winged reptiles, whose whole organism was adapted for flight through the air, appeared in the Jurassic and passed off the stage of existence before the end of the Cretaceous. The bones were hollow, as are those of birds. The sternum, or breastbone, was given a keel for the attachment of the wing muscles. The fifth finger, prodigiously lengthened, was turned backward to support a membrane which was attached to the body and extended to the base of the tail. The other fingers were free, and armed with sharp and delicate claws, as shown in Figures 336 and 337.

These "dragons of the air" varied greatly in size; some were as small as sparrows, while others surpassed in stretch of wing the largest birds of the present day. They may be divided into two groups. The earliest group comprises genera with jaws set with teeth, and with long tails sometimes provided with a rudderlike expansion at the end. In their successors of the later group the tail had become short, and in some of the genera the teeth had disappeared. Among the latest of the flying reptiles was ORNITHOSTOMA (bird beak), the largest creature which ever flew, and whose remains are imbedded in the offshore deposits of the Cretaceous sea which held sway over our western plains. Ornithostoma's spread of wings was twenty feet. Its bones were a marvel of lightness, the entire skeleton, even in its petrified condition, not weighing more than five or six pounds. The sharp beak, a yard long, was toothless and bird-like, as its name suggests

BIRDS. The earliest known birds are found in the Jurassic, and during the remainder of the Mesozoic they contended with the flying reptiles for the empire of the air. The first feathered creatures were very different from the birds of to-day. Their characteristics prove them an offshoot of the dinosaur line of reptiles. ARCHAEOPTERYX (ANCIENT BIRD) (Fig. 338) exhibits a strange mingling of bird and reptile. Like birds, it was fledged with perfect feathers, at least on wings and tail, but it retained the teeth of the reptile, and its long tail was vertebrated, a pair of feathers springing from each joint. Throughout the Jurassic and Cretaceous the remains of birds are far less common than those of flying reptiles, and strata representing hundreds of thousands of years intervene between Archaeopteryx and the next birds of which we know, whose skeletons occur in the Cretaceous beds of western Kansas.

MAMMALS. So far as the entries upon the geological record show, mammals made their advent in a very humble way during the Trias. These earliest of vertebrates which suckle their young were no bigger than young kittens, and their strong affinities with the theromorphs suggest that their ancestors are to be found among some generalized types of that order of reptiles.

During the long ages of the Mesozoic, mammals continued small and few, and were completely dominated by the reptiles. Their remains are exceedingly rare, and consist of minute scattered teeth,--with an occasional detached jaw,--which prove them to have been flesh or insect eaters. In the same way their affinities are seen to be with the lowest of mammals,--the MONOTREMES and MARSUPIALS. The monotremes,--such as the duckbill mole and the spiny ant-eater of Australia, reproduce by means of eggs resembling those of reptiles; the marsupials, such as the opossum and the kangaroo, bring forth their young alive, but in a very immature condition, and carry them for some time after birth in the marsupium, a pouch on the ventral side of the body.

CHAPTER XXI

THE TERTIARY

THE CENOZOIC ERA. The last stages of the Cretaceous are marked by a decadence of the reptiles. By the end of that period the reptilian forms characteristic of the time had become extinct one after another, leaving to represent the class only the types of reptiles which continue to modern times. The day of the ammonite and the belemnite also now drew to a close, and only a few of these cephalopods were left to survive the period. It is therefore at the close of the Cretaceous that the line is drawn which marks the end of the Middle Age of geology and the beginning of the Cenozoic era, the era of modern life,--the Age of Mammals.

In place of the giant reptiles, mammals now become masters of the land, appearing first in generalized types which, during the long ages of the era, gradually evolve to higher forms, more specialized and ever more closely resembling the mammals of the present. In the atmosphere the flying dragons of the Mesozoic give place to birds and bats. In the sea, whales, sharks, and teleost fishes of modern types rule in the stead of huge swimming reptiles. The lower vertebrates, the invertebrates of land and sea, and the plants of field and forest take on a modern aspect, and differ little more from those of to-day than the plants and animals of different countries now differ from one another. From the beginning of the Cenozoic era until now there is a steadily increasing number of species of animals and plants which have continued to exist to the present time.

The Cenozoic era comprises two divisions,--the TERTIARY period and the QUATERNARY period.

In the early days of geology the formations of the entire geological record, so far as it was then known, were divided into three groups,--the PRIMARY, the SECONDARY (now known as the Mesozoic), and the TERTIARY, When the third group was subdivided into two systems, the term Tertiary was retained for the first system of the two, while the term QUATERNARY was used to designate the second.

DIVISIONS OF THE TERTIARY. The formations of the Tertiary are grouped in three divisions,--the PLIOCENE (more recent), the MIOCENE (less recent), and the EOCENE (the dawn of the recent). Each of these epochs is long and complex. Their various sub-divisions are distinguished each by its own peculiar organisms, and the changes of physical geography recorded in their strata. In the rapid view which we are compelled to take we can note only a few of the most conspicuous events of the period.

PHYSICAL GEOGRAPHY OF THE TERTIARY IN EASTERN NORTH AMERICA. The Tertiary rocks of eastern North America are marine deposits and occupy the coastal lowlands of the Atlantic and Gulf states (Fig. 260). In New England, Tertiary beds occur on the island of Martha's Vineyard, but not on the mainland; hence the shore line here stood somewhat farther out than now. From New Jersey southward the earliest Tertiary sands and clays, still unconsolidated, leave only a narrow strip of the edge of the Cretaceous between them and the Triassic and crystalline rocks of the Piedmont oldland; hence the Atlantic shore here stood farther in than now, and at the beginning of the period the present coastal plain was continental delta. A broad belt of Tertiary sea-laid limestones, sandstones, and shales surrounds the Gulf of Mexico and extends northward up the Mississippi embayment to the mouth of the Ohio River; hence the Gulf was then larger than at present, and its waters reached in a broad bay far up the Mississippi valley.

Along the Atlantic coast the Mesozoic peneplain may be traced shoreward to where it disappears from view beneath an unconformable cover of early Tertiary marine strata. The beginning of the Tertiary was therefore marked by a subsidence. The wide erosion surface which at the close of the Mesozoic lay near sea level where the Appalachian Mountains and their neighboring plateaus and uplands now stand was lowered gently along its seaward edge beneath the Tertiary Atlantic to receive a cover of its sediments.

As the period progressed slight oscillations occurred from time to time. Strips of coastal plain were added to the land, and as early as the close of the Miocene the shore lines of the Atlantic and Gulf states had reached well-nigh their present place. Louisiana and Florida were the last areas to emerge wholly from the sea,-- Florida being formed by a broad transverse upwarp of the continental delta at the opening of the Miocene, forming first an island, which afterwards was joined to the mainland.

THE PACIFIC COAST. Tertiary deposits with marine fossils occur along the western foothills of the Sierra Nevadas, and are crumpled among the mountain masses of the Coast Ranges; it is hence inferred that the Great Valley of California was then a border sea, separated from the ocean by a chain of mountainous islands which were upridged into the Coast Ranges at a still later time. Tertiary marine strata are spread over the lower Columbia valley and that of Puget Sound, showing that the Pacific came in broadly there.

THE INTERIOR OF THE WESTERN UNITED STATES. The closing stages of the Mesozoic were marked, as we have seen, by the upheaval of the Rocky Mountains and other western ranges. The bases of the mountains are now skirted by widespread Tertiary deposits, which form the highest strata of the lofty plateaus from the level of whose summits the mountains rise. Like the recent alluvium of the Great Valley of California, these deposits imply low-lying lands when they were laid, and therefore at that time the mountains rose from near sea level. But the height at which the Tertiary strata now stand--five thousand feet above the sea at Denver, and twice that height in the plateaus of southern Utah--proves that the plateaus of which the Tertiary strata form a part have been uplifted during the Cenozoic. During their uplift, warping formed extensive basins both east and west of the Rockies, and in these basins stream-swept and lake-laid waste gathered to depths of hundreds and thousands of feet, as it is accumulating at present in the Great Valley of California and on the river plains of Turkestan. The Tertiary river deposits of the High Plains have already been described. How widespread are these ancient river plains and beds of fresh-water lakes may be seen in the map of Figure 260.

THE BAD LANDS. In several of the western states large areas of Tertiary fresh-water deposits have been dissected to a maze of hills whose steep sides are cut with innumerable ravines. The deposits of these ancient river plains and lake beds are little cemented and because of the dryness of the climate are unprotected by vegetation; hence they are easily carved by the wet-weather rills of scanty and infrequent rains. These waterless, rugged surfaces were named by the early French explorers the BAD LANDS because they were found so difficult to traverse. The strata of the Bad Lands contain vast numbers of the remains of the animals of Tertiary times, and the large amount of barren surface exposed to view makes search for fossils easy and fruitful. These desolate tracts are therefore frequently visited by scientific collecting expeditions.

MOUNTAIN MAKING IN THE TERTIARY. The Tertiary period included epochs when the earth's crust was singularly unquiet. From time to time on all the continents subterranean forces gathered head, and the crust was bent and broken and upridged in lofty mountains.

The Sierra Nevada range was formed, as we have seen, by strata crumpling at the end of the Jurassic. But since that remote time the upfolded mountains had been worn to plains and hilly uplands, the remnants of whose uplifted erosion surfaces may now be traced along the western mountain slopes. Beginning late in the Tertiary, the region was again affected by mountain-making movements. A series of displacements along a profound fault on the eastern side tilted the enormous earth block of the Sierras to the west, lifting its eastern edge to form the lofty crest and giving to the range a steep eastern front and a gentle descent toward the Pacific.

The Coast Ranges also have had a complex history with many vicissitudes. The earliest foldings of their strata belong to the close of the Jurassic, but it was not until the end of the Miocene that the line of mountainous islands and the heavy sediments which had been deposited on their submerged flanks were crushed into a continuous mountain chain. Thick Pliocene beds upon their sides prove that they were depressed to near sea level during the later Tertiary. At the close of the Pliocene the Coast Ranges rose along with the upheaval of the Sierra, and their gradual uplift has continued to the present time.

The numerous north-south ranges of the Great Basin and the Mount Saint Elias range of Alaska were also uptilted during the Tertiary.

During the Tertiary period many of the loftiest mountains of the earth--the Alps, the Apennines, the Pyrenees, the Atlas, the Caucasus, and the Himalayas--received the uplift to which they owe most of their colossal bulk and height, as portions of the Tertiary sea beds now found high upon their flanks attest. In the Himalayas, Tertiary marine limestones occur sixteen thousand five hundred feet above sea level.

VOLCANIC ACTIVITY IN THE TERTIARY. The vast deformations of the Tertiary were accompanied on a corresponding scale by outpourings of lava, the outburst of volcanoes, and the intrusion of molten masses within the crust. In the Sierra Nevadas the Miocene river gravels of the valleys of the western slope, with their placer deposits of gold, were buried beneath streams of lava and beds of tuff. Volcanoes broke forth along the Rocky Mountains and on the plateaus of Utah, New Mexico, and Arizona.

Mount Shasta and the immense volcanic piles of the Cascades date from this period. The mountain basin of the Yellowstone Park was filled to a depth of several thousand feet with tuffs and lavas, the oldest dating as far back as the beginning of the Tertiary. Crandall volcano was reared in the Miocene and the latest eruptions of the Park are far more recent.

THE COLUMBIA AND SNAKE RIVER LAVAS. Still more important is the plateau of lava, more than two hundred thousand square miles in area, extending from the Yellowstone Park to the Cascade Mountains, which has been built from Miocene times to the present.

Over this plateau, which occupies large portions of Idaho, Washington, and Oregon, and extends into northern California and Nevada, the country rock is basaltic lava. For thousands of square miles the surface is a lava plain which meets the boundary mountains as a lake or sea meets a rugged and deeply indented coast. The floods of molten rock spread up the mountain valleys for a score of miles and more, the intervening spurs rising above the lava like long peninsulas, while here and there an isolated peak was left to tower above the inundation like an island off a submerged shore.

The rivers which drain the plateau--the Snake, the Columbia, and their tributaries--have deeply trenched it, yet their canyons, which reach the depth of several thousand feet, have not been worn to the base of the lava except near the margin and where they cut the summits of mountains drowned beneath the flood. Here and there the plateau has been deformed. It has been upbent into great folds, and broken into immense blocks of bedded lava, forming mountain ranges, which run parallel with the Pacific coast line. On the edges of these tilted blocks the thickness of the lava is seen to be fully five thousand feet. The plateau has been built, like that of Iceland, of innumerable overlapping sheets of lava. On the canyon walls they weather back in horizontal terraces and long talus slopes. One may distinguish each successive flow by its dense central portion, often jointed with large vertical columns, and the upper portion with its mass of confused irregular columns and scoriaceous surface. The average thickness of the flows seems to be about seventy-five feet.

The plateau was long in building. Between the layers are found in places old soil beds and forest grounds and the sediments of lakes. Hence the interval between the flows in any locality was sometimes long enough for clays to gather in the lakes which filled depressions in the surface. Again and again the surface of the black basalt was reddened by oxidation and decayed to soil, and forests had time to grow upon it before the succeeding inundation sealed the sediments and soils away beneath a sheet of stone. Near the edges of the lava plain, rivers from the surrounding mountains spread sheets of sand and gravel on the surface of one flow after another. These pervious sands, interbedded with the lava, become the aquifers of artesian wells.

In places the lavas rest on extensive lake deposits, one thousand feet deep, and Miocene in age as their fossils prove. It is to the middle Tertiary, then, that the earliest flows and the largest bulk of the great inundation belong. So ancient are the latest floods in the Columbia basin that they have weathered to a residual yellow clay from thirty to sixty feet in depth and marvelously rich in the mineral substances on which plants feed.

In the Snake River valley the latest lavas are much younger. Their surfaces are so fresh and undecayed that here the effusive eruptions may well have continued to within the period of human history. Low lava domes like those of Iceland mark where last the basalt outwelled and spread far and wide before it chilled (Fig. 341). In places small mounds of scoria show that the eruptions were accompanied to a slight degree by explosions of steam. So fluid was this superheated lava that recent flows have been traced for more than fifty miles.

The rocks underlying the Columbia lavas, where exposed to view, are seen to be cut by numerous great dikes of dense basalt, which mark the fissures through which the molten rock rose to the surface.

The Tertiary included times of widespread and intense volcanic action in other continents as well as in North America. In Europe, Vesuvius and Etna began their career as submarine volcanoes in connection with earth movements which finally lifted Pliocene deposits in Sicily to their present height,--four thousand feet above the sea. Volcanoes broke forth in central France and southern Germany, in Hungary and the Carpathians. Innumerable fissures opened in the crust from the north of Ireland and the western islands of Scotland to the Faroes, Iceland, and even to arctic Greenland; and here great plateaus were built of flows of basalt similar to that of the Columbia River. In India, at the opening of the Tertiary, there had been an outwelling of basalt, flooding to a depth of thousands of feet two hundred thousand square miles of the northwestern part of the peninsula, and similar inundations of lava occurred where are now the table-lands of Abyssinia. From the middle Tertiary on, Asia Minor, Arabia, and Persia were the scenes of volcanic action. In Palestine the rise of the uplands of Judea at the close of the Eocene, and the downfaulting of the Jordan valley were followed by volcanic outbursts. In comparison with the middle Tertiary, the present is a time of volcanic inactivity and repose.

EROSION OF TERTIARY MOUNTAINS AND PLATEAUS. The mountains and plateaus built at various times during the Tertiary and at its commencement have been profoundly carved by erosive agents. The Sierra Nevada Mountains have been dissected on the western slope by such canyons as those of King's River and the Yosemite. Six miles of strata have been denuded from parts of the Wasatch Mountains since their rise at the beginning of the era. From the Colorado plateaus, whose uplift dates from the same time, there have been stripped off ten thousand feet of strata over thousands of square miles, and the colossal canyon of the Colorado has been cut after this great denudation had been mostly accomplished.

On the eastern side of the continent, as we have seen, a broad peneplain had been developed by the close of the Cretaceous. The remnants of this old erosion surface are now found upwarped to various heights in different portions of its area. In southern New England it now stands fifteen hundred feet above the sea in western Massachusetts, declining thence southward and eastward to sea level at the coast. In southwestern Virginia it has been lifted to four thousand feet above the sea. Manifestly this upwarp occurred since the peneplain was formed; it is later than the Mesozoic, and the vast dissection which the peneplain has suffered since its uplift must belong to the successive cycles of Cenozoic time.

Revived by the uplift, the streams of the area trenched it as deeply as its elevation permitted, and reaching grade, opened up wide valleys and new peneplains in the softer rocks. The Connecticut valley is Tertiary in age, and in the weak Triassic sandstones has been widened in places to fifteen miles.

Dating from the same time are the valleys of the Hudson, the Susquehanna, the Delaware, the Potomac, and the Shenandoah.

In Pennsylvania and the states lying to the south the Mesozoic peneplain lies along the summits of the mountain ridges. On the surface of this ancient plain, Tertiary erosion etched out the beautifully regular pattern of the Allegheny mountain ridges and their intervening valleys. The weaker strata of the long, regular folds were eroded into longitudinal valleys, while the hard Paleozoic sandstones, such as the Medina and the Pocono, were left in relief as bold mountain walls whose even crests rise to the common level of the ancient plain. From Virginia far into Alabama the great Appalachian valley was opened to a width in places of fifty miles and more, along a belt of intensely folded and faulted strata where once was the heart of the Appalachian Mountains. In Figure 70 the summit of the Cumberland plateau (ab) marks the level of the Mesozoic peneplain, while the lower erosion levels are Tertiary and Quaternary in age.

LIFE OF THE TERTIARY PERIOD

VEGETATION AND CLIMATE. The highest plants in structure, the DICOTYLS (such as our deciduous forest trees) and the MONOCOTYLS (represented by the palms), were introduced during the Cretaceous. The vegetable kingdom reached its culmination before the animal kingdom, and if the dividing line between the Mesozoic and the Cenozoic were drawn according to the progress of plant life, the Cretaceous instead of the Tertiary would be made the opening period of the modern era.

The plants of the Tertiary belonged, for the most part, to genera now living; but their distribution was very different from that of the flora of to-day. In the earlier Tertiary, palms flourished over northern Europe, and in the northwestern United States grew the magnolia and laurel, along with the walnut, oak, and elm. Even in northern Greenland and in Spitzbergen there were lakes covered with water lilies and surrounded by forests of maples, poplars, limes, the cypress of our southern states, and noble sequoias similar to the "big trees" and redwoods of California. A warm climate like that of the Mesozoic, therefore, prevailed over North America and Europe, extending far toward the pole. In the later Tertiary the climate gradually became cooler. Palms disappeared from Europe, and everywhere the aspect of forests and open lands became more like that of to-day. Grasses became abundant, furnishing a new food for herbivorous animals.

ANIMAL LIFE OF THE TERTIARY. Little needs to be said of the Tertiary invertebrates, so nearly were they like the invertebrates of the present. Even in the Eocene, about five per cent of marine shells were of species still living, and in the Pliocene the proportion had risen to more than one half.

Fishes were of modern types. Teleosts were now abundant. The ocean teemed with sharks, some of them being voracious monsters seventy-five feet and even more in length, with a gape of jaw of six feet, as estimated by the size of their enormous sharp-edged teeth.

Snakes are found for the first time in the early Tertiary. These limbless reptiles, evolved by degeneration from lizardlike ancestors, appeared in nonpoisonous types scarcely to be distinguished from those of the present day.

MAMMALS OF THE EARLY TERTIARY. The fossils of continental deposits of the earliest Eocene show that a marked advance had now been made in the evolution of the Mammalia. The higher mammals had appeared, and henceforth the lower mammals--the monotremes and the marsupials--are reduced to a subordinate place.

These first true mammals were archaic and generalized in structure. Their feet were of the primitive type, with five toes of about equal length. They were also PLANTIGRADES,--that is, they touched the ground with the sole of the entire foot from toe to heel. No foot had yet become adapted to swift running by a decrease in the number of digits and by lifting the heel and sole so that only the toes touch the ground,--a tread called DIGITIGRADE. Nor was there yet any foot like that of the cats, with sharp retractile claws adapted to seizing and tearing the prey. The forearm and the lower leg each had still two separate bones (ulna and radius, fibula and tibia), neither pair having been replaced with a single strong bone, as in the leg of the horse. The teeth also were primitive in type and of full number. The complex heavy grinders of the horse and elephant, the sharp cutting teeth of the carnivores, and the cropping teeth of the grass eaters were all still to come.

Phenacodus is a characteristic genus of the early Eocene, whose species varied in size from that of a bulldog to that of an animal a little larger than a sheep. Its feet were primitive, and their five toes bore nails intermediate in form between a claw and a hoof. The archaic type of teeth indicates that the animal was omnivorous in diet. A cast of the brain cavity shows that, like its associates of the time, its brain was extremely small and nearly smooth, having little more than traces of convolutions.

The long ages of the Eocene and the following epochs of the Tertiary were times of comparatively rapid evolution among the Mammalia. The earliest forms evolved along diverging lines toward the various specialized types of hoofed mammals, rodents, carnivores, proboscidians, the primates, and the other mammalian orders as we know them now. We must describe the Tertiary mammals very briefly, tracing the lines of descent of only a few of the more familiar mammals of the present.

THE HORSE. The pedigree of the horse runs back into the early Eocene through many genera and species to a five-toed, [Footnote: Or, more accurately, with four perfect toes and a rudimentary fifth corresponding to the thumb.] short-legged ancestor little bigger than a cat. Its descendants gradually increased in stature and became better and better adapted to swift running to escape their foes. The leg became longer, and only the tip of the toes struck the ground. The middle toe (digit number three), originally the longest of the five, steadily enlarged, while the remaining digits dwindled and disappeared. The inner digit, corresponding to the great toe and thumb, was the first to go. Next number five, the little finger, was also dropped. By the end of the Eocene a three-toed genus of the horse family had appeared, as large as a sheep. The hoof of digit number three now supported most of the weight, but the slender hoofs of digits two and four were still serviceable. In the Miocene the stature of the ancestors of the horse increased to that of a pony. The feet were still three-toed, but the side hoofs were now mere dewclaws and scarcely touched the ground. The evolution of the family was completed in the Pliocene.

The middle toe was enlarged still more, the side toes were dropped, and the palm and foot bones which supported them were reduced to splints.

While these changes were in progress the radius and ulna of the fore limb became consolidated to a single bone; and in the hind limb the fibula dwindled to a splint, while the tibia was correspondingly enlarged. The molars, also gradually lengthened, and became more and more complex on their grinding surface; the neck became longer; the brain steadily increased in size and its convolutions became more abundant. The evolution of the horse has made for greater fleetness and intelligence.

THE RHINOCEROS AND TAPIR. These animals, which are grouped with the horse among the ODD-TOED (perissodactyl) mammals, are now verging toward extinction. In the rhinoceros, evolution seems to have taken the opposite course from that of the horse. As the animal increased in size it became more clumsy, its limbs became shorter and more massive, and, perhaps because of its great weight, the number of digits were not reduced below the number three. Like other large herbivores, the rhinoceros, too slow to escape its enemies by flight, learned to withstand them. It developed as its means of defense a nasal horn.

Peculiar offshoots of the line appeared at various times in the Tertiary. A rhinoceros, semiaquatic in habits, with curved tusks, resembling in aspect the hippopotamus, lived along the water courses of the plains east of the Rockies, and its bones are now found by the thousands in the Miocene of Kansas. Another developed along a line parallel to that of the horse, and herds of these light-limbed and swift-footed running rhinoceroses ranged the Great Plains from the Dakotas southward.

The tapirs are an ancient family which has changed but little since it separated from the other perissodactyl stocks in the early Tertiary. At present, tapirs are found only in South America and southern Asia,--a remarkable distribution which we could not explain were it not that the geological record shows that during Tertiary times tapirs ranged throughout the northern hemisphere, making their way to South America late in that period. During the Pleistocene they became extinct over all the intervening lands between the widely separated regions where now they live. The geographic distribution of animals, as well as their relationships and origins, can be understood only through a study of their geological history.

THE PROBOSCIDIANS. This unique order of hoofed mammals, of which the elephant is the sole survivor, began, so far as known, in the Eocene, in Egypt, with a piglike ancestor the size of a small horse, with cheek teeth like the Mastodon's, but wanting both trunk and tusks. A proboscidian came next with four short tusks, and in the Miocene there followed a Mastodon (Fig. 346) armed with two pairs of long, straight tusks on which rested a flexible proboscis.

The DINOTHERE was a curious offshoot of the line, which developed in the Miocene in Europe. In this immense proboscidian, whose skull was three feet long, the upper pair of tusks had disappeared, and those of the lower jaw were bent down with a backward curve in walrus fashion.

In the true ELEPHANTS, which do not appear until near the close of the Tertiary, the lower jaw loses its tusks and the grinding teeth become exceedingly complex in structure. The grinding teeth of the mastodon had long roots and low crowns crossed by four or five peaked enameled ridges. In the teeth of the true elephants the crown has become deep, and the ridges of enamel have changed to numerous upright, platelike folds, their interspaces filled with cement. The two genera--Mastodon and Elephant--are connected by species whose teeth are intermediate in pattern. The proboscidians culminated in the Pliocene, when some of the giant elephants reached a height of fourteen feet.

THE ARTIODACTYLS comprise the hoofed Mammalia which have an even number of toes, such as cattle, sheep, and swine. Like the perissodactyls, they are descended from the primitive five-toed plantigrade mammals of the lowest Eocene. In their evolution, digit number one was first dropped, and the middle pair became larger and more massive, while the side digits, numbers two and five, became shorter, weaker, and less serviceable. The FOUR-TOED ARTIODACTYLS culminated in the Tertiary; at present they are represented only by the hippopotamus and the hog. Along the main line of the evolution of the artiodactyls the side toes, digits two and five, disappeared, leaving as proof that they once existed the corresponding bones of palm and sole as splints. The TWO-TOED ARTIODACTYLS, such as the camels, deer, cattle, and sheep, are now the leading types of the herbivores.

SWINE AND PECCARIES are two branches of a common stock, the first developing in the Old World and the second in the New. In the Miocene a noticeable offshoot of the line was a gigantic piglike brute, a root eater, with a skull a yard in length, whose remains are now found in Colorado and South Dakota.

CAMELS AND LLAMAS. The line of camels and llamas developed in North America, where the successive changes from an early Eocene ancestor, no larger than a rabbit, are traced step by step to the present forms, as clearly as is the evolution of the horse. In the late Miocene some of the ancestral forms migrated to the Old World by way of a land connection where Bering Strait now is, and there gave rise to the camels and dromedaries. Others migrated into South America, which had now been connected with our own continent, and these developed into the llamas and guanacos, while those of the race which remained in North America became extinct during the Pleistocene.

Some peculiar branches of the camel stem appeared in North America. In the Pliocene arose a llama with the long neck and limbs of a giraffe, whose food was cropped from the leaves and branches of trees. Far more generalized in structure was the Oreodon, an animal related to the camels, but with distinct affinities also with other lines, such as those of the hog and deer. These curious creatures were much like the peccary in appearance, except for their long tails. In the middle Eocene they roamed in vast herds from Oregon to Kansas and Nebraska.

THE RUMINANTS. This division of the artiodactyls includes antelopes, deer, oxen, bison, sheep, and goats,--all of which belong to a common stock which took its rise in Europe in the upper Eocene from ancestral forms akin to those of the camels. In the Miocene the evolution of the two-toed artiodactyl foot was well-nigh completed. Bonelike growths appeared on the head, and the two groups of the ruminants became specialized,--the deer with bony antlers, shed and renewed each year, and the ruminants with hollow horns, whose two bony knobs upon the skull are covered with permanent, pointed, horny sheaths.

The ruminants evolved in the Old World, and it was not until the later Miocene that the ancestors of the antelope and of some deer found their way to North America. Mountain sheep and goats, the bison and most of the deer, did not arrive until after the close of the Tertiary, and sheep and oxen were introduced by man.

The hoofed mammals of the Tertiary included many offshoots from the main lines which we have traced. Among them were a number of genera of clumsy, ponderous brutes, some almost elephantine in their bulk.

THE CARNIVORES. The ancestral lines of the families of the flesh eaters--such as the cats (lions, tigers, etc.), the bears, the hyenas, and the dogs (including wolves and foxes)--converge in the creodonts of the early Eocene,--an order so generalized that it had affinities not only with the carnivores but also with the insect eaters, the marsupials, and the hoofed mammals as well. From these primitive flesh eaters, with small and simple brains, numerous small teeth, and plantigrade tread, the different families of the carnivores of the present have slowly evolved.

DOGS AND BEARS. The dog family diverged from the creodonts late in the Eocene, and divided into two branches, one of which evolved the wolves and the other the foxes. An offshoot gave rise to the family of the bears, and so closely do these two families, now wide apart, approach as we trace them back in Tertiary times that the Amphicyon, a genus doglike in its teeth and bearlike in other structures, is referred by some to the dog and by others to the bear family. The well-known plantigrade tread of bears is a primitive characteristic which has survived from their creodont ancestry.

CATS. The family of the cats, the most highly specialized of all the carnivores, divided in the Tertiary into two main branches. One, the saber-tooth tigers (Fig. 351), which takes its name from their long, saberlike, sharp-edged upper canine teeth, evolved a succession of genera and species, among them some of the most destructive beasts of prey which ever scourged the earth. They were masters of the entire northern hemisphere during the middle Tertiary, but in Europe during the Pliocene they declined, from unknown causes, and gave place to the other branch of cats,--which includes the lions, tigers, and leopards. In the Americas the saber-tooth tigers long survived the epoch.

MARINE MAMMALS. The carnivorous mammals of the sea--whales, seals, walruses, etc.--seem to have been derived from some of the creodonts of the early Tertiary by adaptation to aquatic life. Whales evolved from some land ancestry at a very early date in the Tertiary; in the marine deposits of the Eocene are found the bones of the Zeuglodon, a whalelike creature seventy feet in length.

PRIMATES. This order, which includes lemurs, monkeys, apes, and man, seems to have sprung from a creodont or insectivorous ancestry in the lower Eocene. Lemur-like types, with small, smooth brains, were abundant in the United States in the early Tertiary, but no primates have been found here in the middle Tertiary and later strata. In Europe true monkeys were introduced in the Miocene, and were abundant until the close of the Tertiary, when they were driven from the continent by the increasing cold.

ADVANCE OF THE MAMMALIA DURING THE TERTIARY. During the several millions of years comprised in Tertiary time the mammals evolved from the lowly, simple types which tenanted the earth at the beginning of the period, into the many kinds of highly specialized mammals of the Pleistocene and the present, each with the various structures of the body adapted to its own peculiar mode of life. The swift feet of the horse, the horns of cattle and the antlers of the deer, the lion's claws and teeth, the long incisors of the beaver, the proboscis of the elephant, were all developed in Tertiary times. In especial the brain of the Tertiary mammals constantly grew larger relatively to the size of body, and the higher portion of the brain--the cerebral lobes--increased in size in comparison with the cerebellum. Some of the hoofed mammals now have a brain eight or ten times the size of that of their early Tertiary predecessors of equal bulk. Nor can we doubt that along with the increasing size of brain went a corresponding increase in the keenness of the senses, in activity and vigor, and in intelligence.

CHAPTER XXII

THE QUATERNARY

The last period of geological history, the Quaternary, may be said to have begun when all, or nearly all, living species of mollusks and most of the existing mammals had appeared.

It is divided into two great epochs. The first, the Pleistocene or Glacial epoch, is marked off from the Tertiary by the occupation of the northern parts of North America and Europe by vast ice sheets; the second, the Recent epoch, began with the disappearance of the ice sheets from these continents, and merges into the present time.

THE PLEISTOCENE EPOCH

We now come to an episode of unusual interest, so different was it from most of the preceding epochs and from the present, and so largely has it influenced the conditions of man's life.

The records of the Glacial epoch are so plain and full that we are compelled to believe what otherwise would seem almost incredible, --that following the mild climate of the Tertiary came a succession of ages when ice fields, like that of Greenland, shrouded the northern parts of North America and Europe and extended far into temperate latitudes.

THE DRIFT. Our studies of glaciers have prepared us to decipher and interpret the history of the Glacial epoch, as it is recorded in the surface deposits known as the drift. Over most of Canada and the northern states this familiar formation is exposed to view in nearly all cuttings which pass below the surface soil. The drift includes two distinct classes of deposits,--the unstratified drift laid down by glacier ice, and the stratified drift spread by glacier waters.

The materials of the drift are in any given place in part unlike the rock on which it rests. They cannot be derived from the underlying rock by weathering, but have been brought from elsewhere. Thus where a region is underlain by sedimentary rocks, as is the drift-covered area from the Hudson River to the Missouri, the drift contains not only fragments of limestone, sandstone, and shale of local derivation, but also pebbles of many igneous and metamorphic rocks, such as granites, gneisses, schists, dike rocks, quartzites, and the quartz of mineral veins, whose nearest source is the Archean area of Canada and the states of our northern border. The drift received its name when it was supposed that the formation had been drifted by floods and icebergs from outside sources,--a theory long since abandoned.

The distribution also of the drift points clearly to its peculiar origin. Within the limits of the glaciated area it covers the country without regard to the relief, mantling with its debris not only lowlands and valleys but also highlands and mountain slopes.

The boundary of the drift is equally independent of the relief of the land, crossing hills and plains impartially, unlike water-laid deposits, whose margins, unless subsequently deformed, are horizontal. The boundary of the drift is strikingly lobate also, bending outward in broad, convex curves, where there are no natural barriers in the topography of the country to set it such a limit. Under these conditions such a lobate margin cannot belong to deposits of rivers, lakes, or ocean, but is precisely that which would mark the edge of a continental glacier which deployed in broad tongues of ice.

THE ROCK SURFACE UNDERLYING THE DRIFT. Over much of its area the drift rests on firm, fresh rock, showing that both the preglacial mantle of residual waste and the partially decomposed and broken rock beneath it have been swept away. The underlying rock, especially if massive, hard, and of a fine grain, has often been ground down to a smooth surface and rubbed to a polish as perfect as that seen on the rock beside an Alpine glacier where the ice has recently melted back. Frequently it has been worn to the smooth, rounded hummocks known as roches moutonnees, and even rocky hills have been thus smoothed to flowing outlines like roches moutonnees on a gigantic scale. The rock pavement beneath the drift is also marked by long, straight, parallel scorings, varying in size from deep grooves to fine striae as delicate as the hair lines cut by an engraver's needle. Where the rock is soft or closely jointed it is often shattered to a depth of several feet beneath the drift, while stony clay has been thrust in among the fragments into which the rock is broken.

In the presence of these glaciated surfaces we cannot doubt that the area of the drift has been overridden by vast sheets of ice which, in their steady flow, rasped and scored the rock bed beneath by means of the stones with which their basal layers were inset, and in places plucked and shattered it.

TILL. The unstratified portion of the drift consists chiefly of sheets of dense, stony clay called till, which clearly are the ground moraines of ancient continental glaciers. Till is an unsorted mixture of materials of all sizes, from fine clay and sand, gravel, pebbles, and cobblestones, to large bowlders. The stones of the till are of many kinds, some having been plucked from the bed rock of the locality where they are found, and others having been brought from outside and often distant places. Land ice is the only agent known which can spread unstratified material in such extensive sheets.

The FINE MATERIAL of the till comes from two different sources. In part it is derived from old residual clays, which in the making had been leached of the lime and other soluble ingredients of the rock from which they weathered. In part it consists of sound rock ground fine; a drop of acid on fresh, clayey till often proves by brisk effervescence that the till contains much undecayed limestone flour. The ice sheet, therefore, both scraped up the mantle of long-weathered waste which covered the country before its coming, and also ground heavily upon the sound rock underneath, and crushed and wore to rock flour the fragments which it carried.

The color of unweathered till depends on that of the materials of which it is composed. Where red sandstones have contributed largely to its making, as over the Triassic sandstones of the eastern states and the Algonkian sandstones about Lake Superior, the drift is reddish. When derived in part from coaly shales, as over many outcrops of the Pennsylvanian, it may when moist be almost black. Fresh till is normally a dull gray or bluish, so largely is it made up of the grindings of unoxidized rocks of these common colors.

Except where composed chiefly of sand or coarser stuff, unweathered till is often exceedingly dense. Can you suggest by what means it has been thus compacted? Did the ice fields of the Glacial epoch bear heavy surface moraines like the medial and lateral moraines of valley glaciers? Where was the greater part of the load of these ice fields carried, judging from what you know of the glaciers of Greenland?

BOWLDERS OF THE DRIFT. The pebbles and bowlders of the drift are in part stream gravels, bowlders of weathering, and other coarse rock waste picked up from the surface of the country by the advancing ice, and in part are fragments plucked from ledges of sound rock after the mantle of waste had been removed. Many of the stones of the till are dressed as only glacier ice can do; their sharp edges have been blunted and their sides faceted and scored.

We may easily find all stages of this process represented among the pebbles of the till. Some are little worn, even on their edges; some are planed and scored on one side only; while some in their long journey have been ground down to many facets and have lost much of their original bulk. Evidently the ice played fast and loose with a stone carried in its basal layers, now holding it fast and rubbing it against the rock beneath, now loosening its grasp and allowing the stone to turn.

Bowlders of the drift are sometimes found on higher ground than their parent ledges. Thus bowlders have been left on the sides of Mount Katahdin, Maine, which were plucked from limestone ledges twelve miles distant and three thousand feet lower than their resting place. In other cases stones have been carried over mountain ranges, as in Vermont, where pebbles of Burlington red sandstone were dragged over the Green Mountains, three thousand feet in height, and left in the Connecticut valley sixty miles away. No other geological agent than glacier ice could do this work.

The bowlders of the drift are often large. Bowlders ten and twenty feet in diameter are not uncommon, and some are known whose diameter exceeds fifty feet. As a rule the average size of bowlders decreases with increasing distance from their sources. Why?

TILL PLAINS. The surface of the drift, where left in its initial state, also displays clear proof of its glacial origin. Over large areas it is spread in level plains of till, perhaps bowlder- dotted, similar to the plains of stony clay left in Spitzbergen by the recent retreat of some of the glaciers of that island. In places the unstratified drift is heaped in hills of various kinds, which we will now describe.

DRUMLINS. Drumlins are smooth, rounded hills composed of till, elliptical in base, and having their longer axes parallel to the movement of the ice as shown by glacial scorings. They crowd certain districts in central New York and in southern Wisconsin, where they may be counted by the thousands. Among the numerous drumlins about Boston is historic Bunker Hill.

Drumlins are made of ground moraine. They were accumulated and given shape beneath the overriding ice, much as are sand bars in a river, or in some instances were carved, like roches moutonnees, by an ice sheet out of the till left by an earlier ice invasion.

TERMINAL MORAINES. The glaciated area is crossed by belts of thickened drift, often a mile or two, and sometimes even ten miles and more, in breadth, which lie transverse to the movement of the ice and clearly are the terminal moraines of ancient ice sheets, marking either the limit of their farthest advance or pauses in their general retreat.

The surface of these moraines is a jumble of elevations and depressions, which vary from low, gentle swells and shallow sags to sharp hills, a hundred feet or so in height, and deep, steep- sided hollows. Such tumultuous hills and hummocks, set with depressions of all shapes, which usually are without outlet and are often occupied by marshes, ponds, and lakes, surely cannot be the work of running water. The hills are heaps of drift, lodged beneath the ice edge or piled along its front. The basins were left among the tangle of morainic knolls and ridges as the margin of the ice moved back and forth. Some bowl-shaped basins were made by the melting of a mass of ice left behind by the retreating glacier and buried in its debris.

THE STRATIFIED DRIFT. Like modern glaciers the ice sheets of the Pleistocene were ever being converted into water about their margins. Their limits on the land were the lines where their onward flow was just balanced by melting and evaporation. On the surface of the ice along the marginal zone, rivulets no doubt flowed in summer, and found their way through crevasses to the interior of the glacier or to the ground. Subglacial streams, like those of the Malaspina glacier, issued from tunnels in the ice, and water ran along the melting ice front as it is seen to do about the glacier tongues of Greenland. All these glacier waters flowed away down the chief drainage channels in swollen rivers loaded with glacial waste.

It is not unexpected therefore that there are found, over all the country where the melting ice retreated, deposits made of the same materials as the till, but sorted and stratified by running water. Some of these were deposited behind the ice front in ice-walled channels, some at the edge of the glaciers by issuing streams, and others were spread to long distances in front of the ice edge by glacial waters as they flowed away.

ESKERS are narrow, winding ridges of stratified sand and gravel whose general course lies parallel with the movement of the glacier. These ridges, though evidently laid by running water, do not follow lines of continuous descent, but may be found to cross river valleys and ascend their sides. Hence the streams by which eskers were laid did not flow unconfined upon the surface of the ground. We may infer that eskers were deposited in the tunnels and ice-walled gorges of glacial streams before they issued from the ice front.

KAMES are sand and gravel knolls, associated for the most part with terminal moraines, and heaped by glacial waters along the margin of the ice.

KAME TERRACES are hummocky embankments of stratified drift sometimes found in rugged regions along the sides of valleys. In these valleys long tongues of glacier ice lay slowly melting. Glacial waters took their way between the edges of the glaciers and the hillside, and here deposited sand and gravel in rude terraces.

Outwash plains are plains of sand and gravel which frequently border terminal moraines on their outward face, and were spread evidently by outwash from the melting ice. Outwash plains are sometimes pitted by bowl-shaped basins where ice blocks were left buried in the sand by the retreating glacier.

Valley trains are deposits of stratified drift with which river valleys have been aggraded. Valleys leading outward from the ice front were flooded by glacial waters and were filled often to great depths with trains of stream-swept drift. Since the disappearance of the ice these glacial flood plains have been dissected by the shrunken rivers of recent times and left on either side the valley in high terraces. Valley trains head in morainic plains, and their material grows finer down valley and coarser toward their sources. Their gradient is commonly greater than that of the present rivers.

THE EXTENT OF THE DRIFT. The extent of the drift of North America and its southern limits are best seen in Figure 359. Its area is reckoned at about four million square miles. The ice fields which once covered so much of our continent were all together ten times as large as the inland ice of Greenland, and about equal to the enormous ice cap which now covers the antarctic regions.

The ice field of Europe was much smaller, measuring about seven hundred and seventy thousand square miles.

CENTERS OF DISPERSION. The direction of the movement of the ice is recorded plainly in the scorings of the rock surface, in the shapes of glaciated hills, in the axes of drumlins and eskers, and in trains of bowlders, when the ledges from which they were plucked can be discovered. In these ways it has been proved that in North America there were three centers where ice gathered to the greatest depth, and from which it flowed in all directions outward. There were thus three vast ice fields,--one the Cordilleran, which lay upon the Cordilleras of British America; one the Keewatin, which flowed out from the province of Keewatin, west of Hudson Bay; and one the LABRADOR ice field, whose center of dispersion was on the highlands of the peninsula of Labrador. As shown in Figure 359, the western ice field extended but a short way beyond the eastern foothills of the Rocky Mountains, where perhaps it met the far-traveled ice from the great central field. The Keewatin and the Labrador ice fields flowed farthest toward the south, and in the Mississippi valley the one reached the mouth of the Missouri and the other nearly to the mouth of the Ohio. In Minnesota and Wisconsin and northward they merged in one vast field.

The thickness of the ice was so great that it buried the highest mountains of eastern North America, as is proved by the transported bowlders which have been found upon their summits. If the land then stood at its present height above sea level, and if the average slope of the ice were no more than ten feet to the mile,--a slope so gentle that the eye could not detect it and less than half the slope of the interior of the inland ice of Greenland,--the ice plateaus about Hudson Bay must have reached a thickness of at least ten thousand feet.

In Europe the Scandinavian plateau was the chief center of dispersion. At the time of greatest glaciation a continuous field of ice extended from the Ural Mountains to the Atlantic, where, off the coasts of Norway and the British Isles, it met the sea in an unbroken ice wall. On the south it reached to southern England, Belgium, and central Germany, and deployed on the eastern plains in wide lobes over Poland and central Russia (Fig. 360).

At the same time the Alps supported giant glaciers many times the size of the surviving glaciers of to-day, and a piedmont glacier covered the plains of northern Switzerland.

THE THICKNESS OF THE DRIFT. The drift is far from uniform in thickness. It is comparatively thin and scanty over the Laurentian highlands and the rugged regions of New England, while from southern New York and Ontario westward over the Mississippi valley, and on the great western plains of Canada, it exceeds an average of one hundred feet over wide areas, and in places has five and six times that thickness. It was to this marginal belt that the ice sheets brought their loads, while northwards, nearer the centers of dispersion, erosion was excessive and deposition slight.

SUCCESSIVE ICE INVASIONS AND THEIR DRIFT SHEETS. Recent studies of the drift prove that it does not consist of one indivisible formation, but includes a number of distinct drift sheets, each with its own peculiar features. The Pleistocene epoch consisted, therefore, of several glacial stages,-- during each of which the ice advanced far southward,--together with the intervening interglacial stages when, under a milder climate, the ice melted back toward its sources or wholly disappeared.

The evidences of such interglacial stages, and the means by which the different drift sheets are told apart, are illustrated in Figure 361. Here the country from N to S is wholly covered by drift, but the drift from N to m is so unlike that from m to S that we may believe it the product of a distinct ice invasion and deposited during another and far later glacial stage. The former drift is very young, for its drainage is as yet immature, and there are many lakes and marshes upon its surface; the latter is far older, for its surface has been thoroughly dissected by its streams. The former is but slightly weathered, while the latter is so old that it is deeply reddened by oxidation and is leached of its soluble ingredients such as lime. The younger drift is bordered by a distinct terminal moraine, while the margin of the older drift is not thus marked. Moreover, the two drift sheets are somewhat unlike in composition, and the different proportion of pebbles of the various kinds of rocks which they contain shows that their respective glaciers followed different tracks and gathered their loads from different regions. Again, in places beneath the younger drift there is found the buried land surface of an older drift with old soils, forest grounds, and vegetable deposits, containing the remains of animals and plants, which tell of the climate of the interglacial stage in which they lived.

By such differences as these the following drift sheets have been made out in America, and similar subdivisions have been recognized in Europe.

5 The Wisconsin formation 4 The Iowan formation 3 The Illinoian formation 2 The Kansan formation 1 The pre-Kansan or Jerseyan formation

In New Jersey and Pennsylvania the edge of a deeply weathered and eroded drift sheet, the Jerseyan, extends beyond the limits of a much younger overlying drift. It may be the equivalent of a deep- buried basal drift sheet found in the Mississippi valley beneath the Kansan and parted from it by peat, old soil, and gravel beds.

The two succeeding stages mark the greatest snowfall of the Glacial epoch. In Kansan times the Keewatin ice field slowly grew southward until it reached fifteen hundred miles from its center of dispersion and extended from the Arctic Ocean to northeastern Kansas. In the Illinoian stage the Labrador ice field stretched from Hudson Straits nearly to the Ohio River in Illinois. In the Iowan and the Wisconsin, the closing stages of the Glacial epoch, the readvancing ice fields fell far short of their former limits in the Mississippi valley, but in the eastern states the Labrador ice field during Wisconsin times overrode for the most part all earlier deposits, and, covering New England, probably met the ocean in a continuous wall of ice which set its bergs afloat from Massachusetts to northern Labrador.

We select for detailed description the Kansan and the Wisconsin formations as representatives, the one of the older and the other of the younger drift sheets.

THE KANSAN FORMATION. The Kansan drift consists for the most part of a sheet of clayey till carrying smaller bowlders than the later drift. Few traces of drumlins, kames, or terminal moraines are found upon the Kansan drift, and where thick enough to mask the preexisting surface, it seems to have been spread originally in level plains of till.

The initial Kansan plain has been worn by running water until there are now left only isolated patches and the narrow strips and crests of the divides, which still rise to the ancient level. The valleys of the larger streams have been opened wide. Their well- developed tributaries have carved nearly the entire plain to valley slopes (Figs. 50 B, and 59). The lakes and marshes which once marked the infancy of the region have long since been effaced. The drift is also deeply weathered. The till, originally blue in color, has been yellowed by oxidation to a depth of ten and twenty feet and even more, and its surface is sometimes rusted to terra-cotta red. To a somewhat less depth it has been leached of its lime and other soluble ingredients. In the weathered zone its pebbles, especially where the till is loose in texture, are sometimes so rotted that granites may be crumbled with the fingers. The Kansan drift is therefore old.

THE WISCONSIN FORMATION. The Wisconsin drift sheet is but little weathered and eroded, and therefore is extremely young. Oxidation has effected it but slightly, and lime and other soluble plant foods remain undissolved even at the grass roots. Its river systems are still in their infancy (Fig. 50, A). Swamps and peat bogs are abundant on its undrained surface, and to this drift sheet belong the lake lands of our northern states and of the Laurentian peneplain of Canada.

The lake basins of the Wisconsin drift are of several different classes. Many are shallow sags in the ground moraine. Still more numerous are the lakes set in hollows among the hills of the terminal moraines; such as the thousands of lakelets of eastern Massachusetts. Indeed, the terminal moraines of the Wisconsin drift may often be roughly traced on maps by means of belts of lakes and ponds. Some lakes are due to the blockade of ancient valleys by morainic delms, and this class includes many of the lakes of the Adirondacks, the mountain regions of New England, and the Laurentian area. Still other lakes rest in rock basins scooped out by glaciers. In many cases lakes are due to more than one cause, as where preglacial valleys have both been basined by the ice and blockaded by its moraines. The Finger lakes of New York, for example, occupy such glacial troughs.

Massive TERMINAL MORAINES, which mark the farthest limits to which the Wisconsin ice advanced, have been traced from Cape Cod and the islands south of New England, across the Appalachians and the Mississippi valley, through the Dakotas, and far to the north over the plains of British America. Where the ice halted for a time in its general retreat, it left RECESSIONAL MORAINES, as this variety of the terminal moraine is called. The moraines of the Wisconsin drift lie upon the country like great festoons, each series of concentric loops marking the utmost advance of broad lobes of the ice margin and the various pauses in their recession.

Behind the terminal moraines lie wide till plains, in places studded thickly with drumlins, or ridged with an occasional esker. Great outwash plains of sand and gravel lie in front of the moraine belts, and long valley trains of coarse gravels tell of the swift and powerful rivers of the time.

THE LOESS OF THE MISSISSIPPI VALLEY. A yellow earth, quite like the loess of China, is laid broadly as a surface deposit over the Mississippi valley from eastern Nebraska to Ohio outside the boundaries of the Iowan and the Wisconsin drift. Much of the loess was deposited in Iowan times. It is younger than the earlier drift sheets, for it overlies their weathered and eroded surfaces. It thickens to the Iowan drift border, but is not found upon that drift. It is older than the Wisconsin, for in many places it passes underneath the Wisconsin terminal moraines. In part the loess seems to have been washed from glacial waste and spread in sluggish glacial waters, and in part to have been distributed by the wind from plains of aggrading glacial streams.

THE EFFECTS OF THE ICE INVASIONS ON RIVERS. The repeated ice invasions of the Pleistocene profoundly disarranged the drainage systems of our northern states. In some regions the ancient valleys were completely filled with drift. On the withdrawal of the ice the streams were compelled to find their way, as best they could, over a fresh land surface, where we now find them flowing on the drift in young, narrow channels. But hundreds of feet below the ground the well driller and the prospector for coal and oil discover deep, wide, buried valleys cut in rock,--the channels of preglacial and interglacial streams. In places the ancient valleys were filled with drift to a depth of a hundred feet, and sometimes even to a depth of four hundred and five hundred feet. In such valleys, rivers now flow high above their ancient beds of rock on floors of valley drift. Many of the valleys of our present rivers are but patchworks of preglacial, interglacial, and postglacial courses (Fig. 366). Here the river winds along an ancient valley with gently sloping sides and a wide alluvial floor perhaps a mile or so in width, and there it enters a young, rock-walled gorge, whose rocky bed may be crossed by ledges over which the river plunges in waterfalls and rapids.

In such cases it is possible that the river was pushed to one side of its former valley by a lobe of ice, and compelled to cut a new channel in the adjacent uplands. A section of the valley may have been blockaded with morainic waste, and the lake formed behind the barrier may have found outlet over the country to one side of the ancient drift-filled valley. In some instances it would seem that during the waning of the ice sheets, glacial streams, while confined within walls of stagnant ice, cut down through the ice and incised their channels on the underlying country, in some cases being let down on old river courses, and in other cases excavating gorges in adjacent uplands.

PLEISTOCENE LAKES. Temporary lakes were formed wherever the ice front dammed the natural drainage of the region. Some, held in the minor valleys crossed by ice lobes, were small, and no doubt many were too short-lived to leave lasting records. Others, long held against the northward sloping country by the retreating ice edge, left in their beaches their clayey beds, and their outlet channels permanent evidences of their area and depth. Some of these glacial lakes are thus known to have been larger than any present lake.

Lake Agassiz, named in honor of the author of the theory of continental glaciation, is supposed to have been held by the united front of the Keewatin and the Labrador ice fields as they finally retreated down the valley of the Red River of the North and the drainage basin of Lake Winnipeg. From first to last Lake Agassiz covered a hundred and ten thousand square miles in Manitoba and the adjacent parts of Minnesota and North Dakota,--an area larger than all the Great Lakes combined. It discharged its waters across the divide which held it on the south, and thus excavated the valley of the Minnesota River. The lake bed--a plain of till--was spread smooth and level as a floor with lacustrine silts. Since Lake Agassiz vanished with the melting back of the ice beyond the outlet by the Nelson River into Hudson Bay, there has gathered on its floor a deep humus, rich in the nitrogenous elements so needful for the growth of plants, and it is to this soil that the region owes its well-known fertility.

THE GREAT LAKES. The basins of the Great Lakes are broad preglacial river valleys, warped by movements of the crust still in progress, enlarged by the erosive action of lobes of the continental ice sheets, and blockaded by their drift. The complicated glacial and postglacial history of the lakes is recorded in old strand lines which have been traced at various heights about them, showing their areas and the levels at which their waters stood at different times.

With the retreat of the lobate Wisconsin ice sheet toward the north and east, the southern and western ends of the basins of the Great Lakes were uncovered first; and here, between the receding ice front and the slopes of land which faced it, lakes gathered which increased constantly in size.

The lake which thus came to occupy the western end of the Lake Superior basin discharged over the divide at Duluth down the St. Croix River, as an old outlet channel proves; that which held the southern end of the basin of Lake Michigan sent its overflow across the divide at Chicago via the Illinois River to the Mississippi; the lake which covered the lowlands about the western end of Lake Erie discharged its waters at Fort Wayne into the Wabash River.

The ice still blocked the Mohawk and St. Lawrence valleys on the east, while on the west it had retreated far to the north. The lakes become confluent in wide expanses of water, whose depths and margins, as shown by their old lake beaches, varied at different times with the position of the confining ice and with warpings of the land. These vast water bodies, which at one or more periods were greater than all the Great Lakes combined, discharged at various times across the divide at Chicago, near Syracuse, New York, down the Mohawk valley, and by a channel from Georgian Bay into the Ottawa River. Last of all the present outlet by the St. Lawrence was established.

The beaches of the glacial lakes just mentioned are now far from horizontal. That of the lake which occupied the Ontario basin has an elevation of three hundred and sixty-two feet above tide at the west and of six hundred and seventy-five feet at the northeast, proving here a differential movement of the land since glacial times amounting to more than three hundred feet. The beaches which mark the successive heights of these glacial lakes are not parallel; hence the warping began before the Glacial epoch closed. We have already seen that the canting of the region is still in progress.

THE CHAMPLAIN SUBSIDENCE. As the Glacial epoch approached its end, and the Labrador ice field melted back for the last time to near its source, the land on which the ice had lain in eastern North America was so depressed that the sea now spread far and wide up the St. Lawrence valley. It joined with Lake Ontario, and extending down the Champlain and Hudson valleys, made an island of New England and the maritime provinces of Canada.

The proofs of this subsidence are found in old sea beaches and sea-laid clays resting on Wisconsin till. At Montreal such terraces are found six hundred and twenty feet above sea level, and along Lake Champlain--where the skeleton of a whale was once found among them--at from five hundred to four hundred feet. The heavy delta which the Mohawk River built at its mouth in this arm of the sea now stands something more than three hundred feet above sea level. The clays of the Champlain subsidence pass under water near the mouth of the

Hudson, and in northern New Jersey they occur two hundred feet below tide. In these elevations we have measures of the warping of the region since glacial times.

THE WESTERN UNITED STATES IN GLACIAL TIMES. The western United States was not covered during the Pleistocene by any general ice sheet, but all the high ranges were capped with permanent snow and nourished valley glaciers, often many times the size of the existing glaciers of the Alps. In almost every valley of the Sierras and the Rockies the records of these vanished ice streams may be found in cirques, glacial troughs, roches moutonnecs, and morainic deposits.

It was during the Glacial epoch that Lakes Bonneville and Lahontan were established in the Great Basin, whose climate must then have been much more moist than now.

THE DRIFTLESS AREA. In the upper Mississippi valley there is an area of about ten thousand square miles in southwestern Wisconsin and the adjacent parts of Iowa and Minnesota, which escaped the ice invasions. The rocks are covered with residual clays, the product of long preglacial weathering. The region is an ancient peneplain, uplifted and dissected in late Tertiary times, with mature valleys whose gentle gradients are unbroken by waterfalls and rapids. Thus the driftless area is in strong contrast with the immature drift topography about it, where lakes and waterfalls are common. It is a bit of preglacial landscape, showing the condition of the entire region before the Glacial epoch.

The driftless area lay to one side of the main track of both the Keewatin and the Labrador ice fields, and at the north it was protected by the upland south of Lake Superior, which weakened and retarded the movement of the ice.

South of the driftless area the Mississippi valley was invaded at different times by ice sheets from the west,--the Kansan and the Iowan,--and again by the Illinoian ice sheet from the east. Again and again the Mississippi River was pushed to one side or the other of its path. The ancient channel which it held along the Illinoian ice front has been traced through southeastern Iowa for many miles.

BENEFITS OF GLACIATION. Like the driftless area, the preglacial surface over which the ice advanced seems to have been well dissected after the late Tertiary uplifts, and to have been carved in many places to steep valley slopes and rugged hills. The retreating ice sheets, which left smooth plains and gently rolling country over the wide belt where glacial deposition exceeded glacial erosion, have made travel and transportation easier than they otherwise would have been.

The preglacial subsoils were residual clays and sands, composed of the insoluble elements of the country rock of the locality, with some minglings of its soluble parts still undissolved. The glacial subsoils are made of rocks of many kinds, still undecayed and largely ground to powder. They thus contain an inexhaustible store of the mineral foods of plants, and in a form made easily ready for plant use.

On the preglacial hillsides the humus layer must have been comparatively thin, while the broad glacial plains have gathered deep black soils, rich in carbon and nitrogen taken from the atmosphere. To these soils and subsoils a large part of the wealth and prosperity of the glaciated regions of our country must be attributed.

The ice invasions have also added very largely to the water power of the country. The rivers which in preglacial times were flowing over graded courses for the most part, were pushed from their old valleys and set to flow on higher levels, where they have developed waterfalls and rapids. This power will probably be fully utilized long before the coal beds of the country are exhausted, and will become one of the chief sources of the national wealth.

THE RECENT EPOCH. The deposits laid since glacial times graduate into those now forming along the ocean shores, on lake beds, and in river valleys. Slow and comparatively slight changes, such as the warpings of the region of the Great Lakes, have brought about the geographical conditions of the present. The physical history of the Recent epoch needs here no special mention.

THE LIFE OF THE QUATERNARY

During the entire Quaternary, invertebrates and plants suffered little change in species,--so slowly are these ancient and comparatively simple organisms modified. The Mammalia, on the other hand, have changed much since the beginning of Quaternary time: the various species of the present have been evolved, and some lines have become extinct. These highly organized vertebrates are evidently less stable than are lower types of animals, and respond more rapidly to changes in the environment.

PLEISTOCENE MAMMALS. In the Pleistocene the Mammalia reached their culmination both in size and in variety of forms, and were superior in both these respects to the mammals of to-day. In Pleistocene times in North America there were several species of bison,--one whose widespreading horns were ten feet from tip to tip,--a gigantic moose elk, a giant rodent (Castoroides) five feet long, several species of musk oxen, several species of horses,-- more akin, however, to zebras than to the modern horse,--a huge lion, several saber-tooth tigers, immense edentates of several genera, and largest of all the mastodon and mammoth.

The largest of the edentates was the Megatherium, a. clumsy ground sloth bigger than a rhinoceros. The bones of the Megatherium are extraordinarily massive,--the thigh bone being thrice as thick as that of an elephant,--and the animal seems to have been well able to get its living by overthrowing trees and stripping off their leaves. The Glyptodon was a mailed edentate, eight feet long, resembling the little armadillo. These edentates survived from Tertiary times, and in the warmer stages of the Pleistocene ranged north as far as Ohio and Oregon.

The great proboscidians of the Glacial epoch were about the size of modern elephants, and somewhat smaller than their ancestral species in the Pliocene. The MASTODON ranged over all North America south of Hudson Bay, but had become extinct in the Old World at the end of the Tertiary. The elephants were represented by the MAMMOTH, which roamed in immense herds from our middle states to Alaska, and from Arctic Asia to the Mediterranean and Atlantic.

It is an oft-told story how about a century ago, near the Lena River in Siberia, there was found the body of a mammoth which had been safely preserved in ice for thousands of years, how the flesh was eaten by dogs and bears, and how the eyes and hoofs and portions of the hide were taken with the skeleton to St. Petersburg. Since then several other carcasses of the mammoth, similarly preserved in ice, have been found in the same region,-- one as recently as 1901. We know from these remains that the animal was clothed in a coat of long, coarse hair, with thick brown fur beneath.

THE DISTRIBUTION OF ANIMALS AND PLANTS. The distribution of species in the Glacial epoch was far different from that of the present. In the glacial stages arctic species ranged south into what are now temperate latitudes. The walrus throve along the shores of Virginia and the musk ox grazed in Iowa and Kentucky. In Europe the reindeer and arctic fox reached the Pyrenees. During the Champlain depression arctic shells lived along the shore of the arm of the sea which covered the St. Lawrence valley. In interglacial times of milder climate the arctic fauna-flora retreated, and their places were taken by plants and animals from the south. Peccaries, now found in Texas, ranged into Michigan and New York, while great sloths from South America reached the middle states. Interglacial beds at Toronto, Canada, contain remains of forests of maple, elm, and papaw, with mollusks now living in the Mississippi basin.

What changes in the forests of your region would be brought about, and in what way, if the climate should very gradually grow colder? What changes if it should grow warmer?

On the Alps and the highest summits of the White Mountains of New England are found colonies of arctic species of plants and insects. How did they come to be thus separated from their home beyond the arctic circle by a thousand miles and more of temperate climate impossible to cross?

MAN. Along with the remains of the characteristic animals of the time which are now extinct there have been found in deposits of the Glacial epoch in the Old World relics of Pleistocene MAN, his bones, and articles of his manufacture. In Europe, where they have best been studied, human relics occur chiefly in peat bogs, in loess, in caverns where man made his home, and in high river terraces sometimes eighty and a hundred feet above the present flood plains of the streams.

In order to understand the development of early man, we should know that prehistoric peoples are ranked according to the materials of which their tools were made and the skill shown in their manufacture. There are thus four well-marked stages of human culture preceding the written annals of history:

4 The Iron stage. 3 The Bronze stage. 2 The Neolithic (recent stone) stage. 1 The Paleolithic (ancient stone) stage.

In the Neolithic stage the use of the metals had not yet been learned, but tools of stone were carefully shaped and polished. To this stage the North American Indian belonged at the time of the discovery of the continent. In the Paleolithic stage, stone implements were chipped to rude shapes and left unpolished. This, the lowest state of human culture, has been outgrown by nearly every savage tribe now on earth. A still earlier stage may once have existed, when man had not learned so much as to shape his weapons to his needs, but used chance pebbles and rock splinters in their natural forms; of such a stage, however, we have no evidence.

PALEOLITHIC MAN IN EUROPE. It was to the Paleolithic stage that the earliest men belonged whose relics are found in Europe. They had learned to knock off two-edged flakes from flint pebbles, and to work them into simple weapons. The great discovery had been made that fire could be kindled and made use of, as the charcoal and the stones discolored by heat of their ancient hearths attest. Caves and shelters beneath overhanging cliffs were their homes or camping places. Paleolithic man was a savage of the lowest type, who lived by hunting the wild beasts of the time.

Skeletons found in certain caves in Belgium and France represent perhaps the earliest race yet found in Europe. These short, broad- shouldered men, muscular, with bent knees and stooping gait, low- browed and small of brain, were of little intelligence and yet truly human.

The remains of Pleistocene man are naturally found either in caverns, where they escaped destruction by the ice sheets, or in deposits outside the glaciated area. In both cases it is extremely difficult, or quite impossible, to assign the remains to definite glacial or interglacial times. Their relative age is best told by the fauna with which they are associated. Thus the oldest relics of man are found with the animals of the late Tertiary or early Quaternary, such as a species of hippopotamus and an elephant more ancient than the mammoth. Later in age are the remains found along with the mammoth, cave bear and cave hyena, and other animals of glacial time which are now extinct; while more recent still are those associated with the reindeer, which in the last ice invasion roamed widely with the mammoth over central Europe.

THE CAVES OF SOUTHERN FRANCE. These contain the fullest records of the race, much like the Eskimos in bodily frame, which lived in western Europe at the time of the mammoth and the reindeer. The floors of these caves are covered with a layer of bone fragments, the remains of many meals, and here are found also various articles of handicraft. In this way we know that the savages who made these caves their homes fished with harpoons of bone, and hunted with spears and darts tipped with flint and horn. The larger bones are split for the extraction of the marrow. Among such fragments no split human bones are found; this people, therefore, were not cannibals. Bone needles imply the art of sewing, and therefore the use of clothing, made no doubt of skins; while various ornaments, such as necklaces of shells, show how ancient is the love of personal adornment. Pottery was not yet invented. There is no sign of agriculture. No animals had yet been domesticated; not even man's earliest friend, the dog. Certain implements, perhaps used as the insignia of office, suggest a rude tribal organization and the beginnings of the state. The remains of funeral feasts in front of caverns used as tombs point to a religion and the belief in a life beyond the grave. In the caverns of southern France are found also the beginnings of the arts of painting and of sculpture. With surprising skill these Paleolithic men sketched on bits of ivory the mammoth with his long hair and huge curved tusks, frescoed their cavern walls with pictures of the bison and other animals, and carved reindeer on their dagger heads.

EARLY MAN ON OTHER CONTINENTS. Paleolithic flints curiously like those of western Europe are found also in many regions of the Old World,--in India, Egypt, and Asia Minor,--beneath the earliest vestiges of the civilization of those ancient seats, and sometimes associated with the fauna of the Glacial epoch.

In Java there were found in 1891, in strata early Quaternary or late Pliocene in age, parts of a skeleton of lower grade, if not of greater antiquity, than any human remains now known. Pithecanthropus erectus, as the creature has been named, walked erect, as its thigh bone shows, but the skull and teeth indicate a close affinity with the ape.

In North America there have been reported many finds of human relics in valley trains, loess, old river gravels buried beneath lava flows, and other deposits of supposed glacial age; but in the opinion of some geologists sufficient proof of the existence of man in America in glacial times has not as yet been found.

These finds in North America have been discredited for various reasons. Some were not made by scientific men accustomed to the closest scrutiny of every detail. Some were reported after a number of years, when the circumstances might not be accurately remembered; while in a number of instances it seems possible that the relics might have been worked into glacial deposits by natural causes from the surface.

Man, we may believe, witnessed the great ice fields of Europe, if not of America, and perhaps appeared on earth under the genial climate of preglacial times. Nothing has yet been found of the line of man's supposed descent from the primates of the early Tertiary, with the possible exception of the Java remains just mentioned. The structures of man's body show that he is not descended from any of the existing genera of apes. And although he may not have been exempt from the law of evolution,--that method of creation which has made all life on earth akin,--yet his appearance was an event which in importance ranks with the advent of life upon the planet, and marks a new manifestation of creative energy upon a higher plane. There now appeared intelligence, reason, a moral nature, and a capacity for self-directed progress such as had never been before on earth.

THE RECENT EPOCH. The Glacial epoch ends with the melting of the ice sheets of North America and Europe, and the replacement of the Pleistocene mammalian fauna by present species. How gradually the one epoch shades into the other is seen in the fact that the glaciers which still linger in Norway and Alaska are the lineal descendants or the renewed appearances of the ice fields of glacial times.

Our science cannot foretell whether all traces of the Great Ice Age are to disappear, and the earth is to enjoy again the genial climate of the Tertiary, or whether the present is an interglacial epoch and the northern lands are comparatively soon again to be wrapped in ice.

NEOLITHIC MAN. The wild Paleolithic men vanished from Europe with the wild beasts which they hunted, and their place was taken by tribes, perhaps from Asia, of a higher culture. The remains of Neolithic man are found, much as are those of the North American Indians, upon or near the surface, in burial mounds, in shell heaps (the refuse heaps of their settlements), in peat bogs, caves, recent flood-plain deposits, and in the beds of lakes near shore where they sometimes built their dwellings upon piles.

The successive stages in European culture are well displayed in the peat bogs of Denmark. The lowest layers contain the polished STONE implements of Neolithic man, along with remains of the SCOTCH FIR. Above are OAK trunks with implements of BRONZE, while the higher layers hold iron weapons and the remains of a BEECH forest.

Neolithic man in Europe had learned to make pottery, to spin and weave linen, to hew timbers and build boats, and to grow wheat and barley. The dog, horse, ox, sheep, goat, and hog had been domesticated, and, as these species are not known to have existed before in Europe, it is a fair inference that they were brought by man from another continent of the Old World. Neolithic man knew nothing of the art of extracting the metals from their ores, nor had he a written language.

The Neolithic stage of culture passes by insensible gradations into that of the age of bronze, and thus into the Recent epoch.

In the Recent epoch the progress of man in language, in social organization, in the arts of life, in morals and religion, has left ample records which are for other sciences than ours to read; here, therefore, geology gives place to archaeology and history.

Our brief study of the outlines of geology has given us, it is hoped, some great and lasting good. To conceive a past so different from the present has stimulated the imagination, and to follow the inferences by which the conclusions of our science have been reached has exercised one of the noblest faculties of the mind,--the reason. We have learned to look on nature in new ways: every landscape, every pebble now has a meaning and tells something of its origin and history, while plants and animals have a closer interest since we have traced the long lines of their descent. The narrow horizons of human life have been broken through, and we have caught glimpses of that immeasurable reach of time in which nebulae and suns and planets run their courses. Moreover, we have learned something of that orderly and world-embracing progress by which the once uninhabitable globe has come to be man's well-appointed home, and life appearing in the lowliest forms has steadily developed higher and still higher types. Seeing this process enter human history and lift our race continually to loftier levels, we find reason to believe that the onward, upward movement of the geological past is the manifestation of the same wise Power which makes for righteousness and good and that this unceasing purpose will still lead on to nobler ends.

###

www.ingramcontent.com/pod-product-compliance
Lightning Source LLC
Chambersburg PA
CBHW051800170526
45167CB00005B/1817